nival- of or relating to

MW01485804

Transportation and Revolt

Transportation and Revolt

Pigeons, Mules, Canals, and the Vanishing Geographies
of Subversive Mobility

Jacob Shell

The MIT Press
Cambridge, Massachusetts
London, England

© 2015 Massachusetts Institute of Technology

All rights reserved. No part of this book may be reproduced in any form by any electronic or mechanical means (including photocopying, recording, or information storage and retrieval) without permission in writing from the publisher.

MIT Press books may be purchased at special quantity discounts for business or sales promotional use. For information, please email special_sales@mitpress.mit.edu.

This book was set in ITC Stone Serif Std 9/13pt by Toppan Best-set Premedia Limited, Hong Kong. Printed and bound in the United States of America.

Library of Congress Cataloging-in-Publication Data is available.

ISBN: 978-0-262-02933-9

10 9 8 7 6 5 4 3 2 1

For Ize

Contents

Acknowledgments

While I was writing this book, I was fortunate to cross paths with a great many teachers and advisers who offered fresh, catalyzing ideas. Don Mitchell, Mark Monmonier, Anne Mosher, Tod Rutherford, Susan Millar, Robert Fishman, Susan Christopherson, Holly Dobbins, John Mercer, Bill Kelleher, Chrissy Hosea, and Robert Wilson all provided important encouragement and critiques that helped give form to this project. Other valuable insights and suggestions came from Walther Schoonenberg, Wendy Freer, Peter Kishore Saval, Nikil Saval, Daniel Schlozman, Kafui Attoh, He Wang, Chen-I Kuan, Francis Lévesque, Ray Shill, Richard Lair, Gail Malmgreen, Mike Ruddy, Robert Norris, Piet de Rooy, and Jianmei Ni. Finally, writing the book would have been impossible without the support and inspiration provided by my wife, Ize, my parents, Marc and Susan Shell, my sister, Hanna Rose Shell, and my grandparents, Sophie and Murray Meld and Sophie Shell.

Introduction

During World War I, German soldiers in occupied Belgium were known to shoot at flocks of pigeons overhead, for fear the birds were smuggling the missives of insurgents and spies. At Middelkerke in 1915, soldiers shot a pigeon that had just arrived from nearby Ostend, on the pretext that the pigeon was carrying an enemy communiqué, and the German authorities fined the town one million marks. The burgomaster demanded to see evidence that there was indeed such a letter affixed to the bird's leg, but was refused. At the end of May, in Bruges and Ostend, the German high command, "always haunted by the fear of espionage," ordered that soldiers carry out a "general massacre of carrier pigeons" in the region.[1]

In England near the turn of the twenty-first century, some elderly pigeon keepers recalled comparable scenes, from their youths during World War II, of British government agents' ordering the systematic slaughter of coops of carrier pigeons. Recollected one English woman of her childhood in Shrewsbury:

We used to keep pigeons as pets before the war. They'd come into the house and we gave them all names. When the war began though, the government passed a regulation saying that all pigeons had to be destroyed in case they were used to send messages to the Germans. Our pigeons were taken away from us and put down. My brother and I were terribly upset.[2]

Wrote another Briton, in a letter to the journal *British Archaeology*: "Pigeons were still using the loft until the outbreak of World War II. Then, I remember my father being very upset by a government instruction that all pigeons had to be destroyed in case they were used as carriers, and the openings sealed so that birds could not use them."[3] Belgium's most famous pigeon fancier of the twentieth century, Father Cornelis Koopman, similarly recollected that during World War II, all his pigeons had to be destroyed, on the orders of the Germans.[4] During the war the use of carrier pigeons was prohibited at British and Australian internment camps too, with Japanese-, German-, and Italian-speaking inmates forbidden from approaching birds that had flown onto the compound grounds.[5]

In some regimes, top-down concerns regarding the relationship between a population and its pigeons lingered well after the great world wars of the first half of the twentieth century. For instance, in the USSR during the 1970s, all carrier pigeons had to be registered and their use always reported, so that tight control could be maintained over social communications.[6] The Taliban, seizing power in Afghanistan after the Soviets' failed war there, banned the use of carrier pigeons.[7] During a congress of China's governing party in Beijing in 2012, amateur pigeon racers in the region were prohibited from flying their birds.[8]

Numerous observers have noted the special skill of the carrier pigeon, and its human handlers, in the art of moving papers in secret. Che Guevara, in his 1961 study of effective techniques of guerrilla warfare in countries undergoing social revolution, advocated use of the carrier pigeon as a method of eluding the better-equipped enemy's powers of surveillance and message interception.[9] In the late twentieth and early twenty-first centuries, some militaries, such as those in Spain, France, and China, have maintained their own pigeon messenger services, as this method of delivery remains, as one historian of the pigeon remarks, "impervious to modern eavesdropping devices."[10]

Nor, as other observers have noted, is the carrier pigeon's usefulness limited to the transport of scrolled paper. One medical doctor in London during the 1950s described how

even today, when we have highly efficient and speedy carriers, pigeons are still used in London to transport samples of blood from several hospitals to a central laboratory. The birds fly the blood samples over the congested traffic jams to lofts on the lab-roofs, saving hours of critical time.[11]

The skill of the carrier pigeon as a lightweight freight transporter has not gone unnoticed by smugglers. Gold and gems have been flown from country to country within Asia by way of trained homing birds, and in the twenty-first century diamond miners in South Africa have taped rough-cut stones to the ankles and chests of pigeons and watched them fly off with the illicit cargo in the direction of the garden coops the pigeons know as home.[12] Inmates in South American prisons have been observed receiving into their cells contraband such as handheld phone parts, delivered by way of trained pigeons.[13] In India, anti-terrorist officials have voiced and acted on suspicions that carrier pigeons were carrying data chips from terrorist cell to terrorist cell across South Asia.[14]

Figure 0.1
Police apprehended this carrier pigeon conveying a cell phone into a prison in Pirajuí, Brazil, in 2012.
Source: Fotos cedidas/ Jornal da Cidade.

One writer has noted that pigeons have also acted as narcotics smugglers in South America and on the U.S.-Mexican border—that, indeed, "a border patrolman or a U.S. Customs officer can go crazy trying to catch a sixty-miles-an-hour bandit pigeon whirring five hundred feet up in the air."[15] In American popular culture the pigeon is more often associated with the inner-city street bandit than with the bandit of the peripheral border country. Marlon Brando's harbor miscreant in the film *On the Waterfront* (1954) and Forest Whitaker's mafia retainer in *Ghost Dog* (1999) keep pigeons in makeshift coops on city rooftops. Often, in this genre of storytelling, such pigeons wind up being massacred, in retaliation for the keeper having become a "stool pigeon" who traffics information in the wrong direction.

This book pursues the following line of inquiry: What sorts of carrying technologies have political regimes associated with the movement of weapons, papers, or people for political subversion and revolt? In an era when much information transfer occurs across a wiretappable medium and much transport of goods and people occurs across a mapped network of tracks and checkpoints, what social history of the specter of subversive trafficking—and of the associated political fears this specter has been able to elicit—might help us better understand the retrenchment of an older range of possibilities for human mobility?

Subversive mobility (the word "subversive" can be used in many senses, and here I use it only to mean "threatening to the continued existence of a ruling political regime") is a topic treated in a growing body of scholarship in such fields as social history, political theory, cultural studies, and human geography. Studies of subversive mobility have looked at such themes as the smuggling and clandestine circulation of arms and insurrectionary papers among politically marginalized, radicalizing social groups;[16] the secret routes taken by political fugitives escaping slavery or other forms of political oppression;[17] parallels between the movements of migrant workers and the spread of radical political ideas;[18] and patterns of geographic advance during violent popular movements, such as food riots and agrarian rebellions.[19] What I add to this area of research is an identification and analysis of those transport modes understood, in different times and places, as especially instrumental for the facilitation of politically subversive, clandestine movement—instrumental to the point, even, of discouraging continued top-down political and institutional support for these particular technological capacities for human mobility.

The political theorist James C. Scott's intriguing suggestion, at the beginning of his 1998 *Seeing Like a State*, that, under certain circumstances, modern ruling regimes may select or reject entire organizations of transportation based on a political desire to diminish the ability of a ruled population to mobilize for revolt, provides a conceptual launching point for the examinations that follow here.[20] Another important launching point is the 2001 book *The Many-Headed Hydra*, by the social historians Peter Linebaugh and Marcus Rediker. Linebaugh and Rediker's work focuses on subversive trafficking networks and patterns of physical and political intercourse among many rebellious social groups and anti-imperial political cells throughout the North Atlantic maritime world of the seventeenth, eighteenth, and early nineteenth centuries. Linebaugh and Rediker emphasize the centrality of certain maritime transport spaces—ship hulls, docks, waterfront haunts, canals and bayous—to this oppositional network. The authors tentatively propose that the replacement, in the mid-nineteenth century, of wind power and wood by steam power and iron as the main physical bases of shipping technology represented a ruling-class attempt (often a successful one) at breaking up this subversive network.[21] In other words, Linebaugh and Rediker, much like Scott, put forward the idea that, under certain conditions, ruling regimes may attempt to reconfigure the geographic and technological basis of transportation, with the aim of eroding a ruled population's field of opportunities for rebellious political interaction.

My analytical tendencies differ somewhat from these authors'. Frequently, the discussion here approaches the history of subversive patterns of human mobility, and of the political fears these patterns have elicited, from the standpoint of a related history of unrealized transport schemes—a subject area neither Scott nor Linebaugh and Rediker touch on. In many of the cases I present in this book, the transport infrastructures that ruling regimes refused to build offer us a measure not of those regimes' tightfisted economic sense or impoverished spatial vision but rather of their fear of the populations over which they exercised power. For this reason, much of the book gathers together subject matter normally relegated to planning and transportation history (and sometimes also society and technology studies) with subject matter normally relegated to radical social history and radical historical geography. Some readers might find this combination of thematic elements unusual (but perhaps also stimulating); it is a combination that, I believe, is intrinsically necessary for the argument at hand. It means relatively little to say that in some places during past eras, certain kinds of transport labor were objects of political suspicion. *Most* kinds of labor become objects of political suspicion at some point or another. It means much more to show how in some cases, these suspicions provoked destructive action from ruling elites. This destructive action could be a massacre of implicated beasts of burden, such as carrier pigeons. Or it could be the shelving of otherwise promising proposals to expand on infrastructure systems associated with patterns of mobility that ruling parties found some reason to fear. Chapter 3 presents one example of this second form of destruction, looking at the marginalization and dismissal of numerous inland ship canal schemes in Britain, Ireland, and British North America during the late nineteenth and early twentieth centuries, a period when existing canal networks in these areas were often perceived as having been coopted by the forces of anti-imperial revolt.

In a sense, the account of the carrier pigeon as subversive smuggler acts as a kind of microcosm or "dumb show" for the principal themes at play in this book. The image of German soldiers shooting fearfully at flocks of pigeons in the Belgian sky is especially fitting for my purpose, which is to focus as much on the consequential fears elicited by certain carrying systems as on the carrying systems themselves. Building on this pigeon preamble, chapter 1 takes up the usefulness of another transport animal, the mule, for cargo smuggling and for politically subversive, evasive movement across rough or mountainous terrain, where regular roads cannot easily go. I highlight numerous examples of this particular utilization of mules: in northern Mexico in the 1910s, in American Appalachia in the 1920s and 1930s, in Greece in the late 1940s, in Cuba and Korea in the 1950s. I also argue that, at least in the United States, this special utility of muleteering in aiding guerrilla logistics helps to more fully elucidate a related history of social and political biases directed against mules and mule handlers—biases that, I submit, place in new historical light institutional divestments from the American domestic mule industry in the middle of the twentieth century.

Chapter 2 expands on this framework, building up a kind a conceptual spectrum, or what I refer to as a "color wheel," of other kinds of transportation that are able to access geophysical zones unreachable by roads and tracks. The usefulness of trained elephants for subversive mobility across mud, of camels for subversive mobility across windswept desert sands, and of sled dogs for subversive mobility across zones of snow and ice is, I propose, closely related to the usefulness of mules on jagged terrain and carrier pigeons in the air. Also analogous are the advantages of watercraft to the rebel transporter or logistician; indeed, the parallel between bandit-cameleers and pirate navies was noted by two widely read commentators during the early twentieth century, the American geographer Ellen Semple and the British military officer and writer T. E. Lawrence. In all, then, this chapter presents six forms of transportation, each suited to subversive mobility across a different type of road-resistant geophysical zone: air, jagged rock, mud, windblown sand, ice, and liquid water. The discussion touches on numerous geographic areas: Burma (Myanmar), India, Central and East Africa, Australia, Siberia, and Shanghai, China. In bringing this array of material together, my object is to suggest a more widely applicable theoretical framework for thinking about how subversive transportative possibilities might manifest themselves across geophysical space.

Also introduced and partially explored in these first two chapters is the theme of political fear as a motivating factor in the removal of select transport capacities from the modern world (or at least from many areas of the modern world). The carrier pigeon massacres recorded in Belgium and Britain during the world wars of the twentieth century provide a preliminary illustration of this theme. As I discuss in chapter 2, the comparable destruction of the Chukchi sled dog in far eastern Siberia under Soviet rule provides another example. However, such "transport destruction" is not always so direct that it comes out of the barrel of a gun. It can instead entail the removal of certain kinds of social knowledge: how to pack a mule, how to ride an elephant through muddy forests during a monsoon, how to navigate labyrinthine marshlands by boat at night, and so on. Consciously or not, political and financial leaders can also effectively destroy or curtail a transport method by denying that method a useful niche in the modern world. Such niches might include transportation for flood-time rescues (for which, in appropriate climates, elephants can be uniquely useful, when floods wash away roads), surficial resource exploration and prospecting (for which pack animals in general can be especially valuable), or counter-guerrilla operations in war zones where there are few usable roads (for which mules in particular can prove indispensable).

Similarly, regimes can truncate the social use and spatial extent of transport methods associated with subversive patterns of mobility by dismissing, or shelving, closely studied and economically promising planning schemes to modernize, expand, or fill in "missing links" within these transport methods' associated infrastructure networks. Chapter 3, for instance, points to the British Empire's record of ignoring heavily studied inland ship canal schemes between 1870 and 1920—a period during which rival industrial empires were at work constructing what one canal historian has called the inland "shipping canal era," but during which British elites mostly halted new canal construction or canal modernization projects. In the British realm during this period, canal transport was emphatically stymied in favor of rail transport. This process of transportative "outcompetition" was not technologically preordained or economically inevitable, nor was the demise of British canal building some sort of inescapable consequence of the institutional balance of British political economy during the late nineteenth and early twentieth centuries. Rather, I suggest that Britain's "missing ship canal era" needs to be understood in relation to a set of social biases and political fears peculiar to the British ruling and middle classes during this period—biases against the people who worked the boats along the existing British canal networks, as well as fears that these "canallers" were facilitating the secret shipment of weapons, papers, and people in aid of anti-imperial revolt.

Chapter 4 points to urban deindustrialization as another modern geographic outcome that can, at least in some instances, be better accounted for by taking into consideration a preceding history of subversive transport networks, and of top-down efforts to frustrate real or imagined versions of those networks. This chapter focuses on the interconnected histories of violent class conflict, political fear, freight transport investment, and city planning in and around the Port of New York in the early and mid-twentieth century. I highlight a puzzle in the history of New York City's infrastructural development: why, during the wave of modernist planning efforts and infrastructure projects that dramatically reshaped the city during the middle third of the twentieth century, did none of these planning projects make allowance for, or invest in, urban *freight* transportation? As with the British canals case, I point to a mass of intensively studied but unrealized schemes: plans, especially between the world wars, to build freight subways and stevedoring facilities along the inner-city waterfront. I highlight this body of plans so as to more palpably sketch out the sorts of possibilities for economic activity and physical mobility that were understood as quite feasible during this era but were blocked nonetheless. In querying the social and political context of these plans' marginalization and dismissal, I look at length at the 1919 memoirs of New York's World War I–era Bomb Squad police captain. These memoirs implicated the port's small watercraft pilots in the trafficking and setting off of German- and anarchist-made bombs around the port. Over subsequent years, these memoirs were read by counter-subversion officials and business leaders in New York and Washington. Alongside these memoirs, I also look at the institutional commitments surrounding an influential and at least partly realized economic and infrastructural planning study of the 1920s that, among other things, advocated the sweeping deindustrialization of the urban core at a time when inner-city manufacturing was still economically ascendant. My aim in presenting these texts together is to provide distinct yet related windows into the sorts of economic biases, social preoccupations, and political anxieties that were motivating elite thinking in New York at the time. In turn, my aim is also to place subsequent *planning* biases, which aggressively enshrined speculative office real estate as the linchpin of the city's economy, in a new historical light.

At first glance, chapter 4 may seem to present something of a shift in gears compared with the previous three chapters. It focuses on a single place, New York City, rather than on entire nations or entire empires, as much of the discussion in chapters 1, 2, and 3 does. This shift in scalar focus deserves some explanation. What the New York example serves to demonstrate is that the history of city planning—and by extension of urban transport— has been shaped by radical social history in ways that planning historians have mostly overlooked. Although from one point of view my argument in this chapter can speak only to what transpired in this particular place, it is significant that from a city planning perspective, New York was in many ways looked to as a model for other American cities, and for some non-American cities as well, over the course of the twentieth century. Thus, New York City's associated corpus of urban planning schemes, both built and unbuilt, presents an important foundation for examining a wider history of urban freight transport—and by extension of urban industrialization and deindustrialization. It is at least plausible that had a different planning paradigm emerged from New York's industrial heyday, this paradigm would have influenced twentieth-century urban planning theories and policies on a wider scale. I submit this case study, then, both as an in-depth argument about a single place and as an effort to cast new light on directions taken, and not taken, by practitioners of urban planning and transport development over the course of the twentieth century.

Ruling regimes are not always explicit in expressing their fears that a given mode of transportation is proving unacceptably instrumental for the mobility of insurgents, political fugitives, and subversive smugglers. Throughout the book, I employ a variety of strategies to tease out, from written records, political understandings of, and wider cultural attitudes toward, the methods of transportation under consideration. One such strategy is the analysis of police or military officials' written commentaries on the logistics (or what such officials perceived to be the logistics) of subversive mobility during revolts. Another strategy is close scrutiny of the literary tropes and evocative associations employed by travel writers, novelists, and other kinds of storytellers in their descriptive treatment of certain classes of transport and transport labor. At times, I appeal to a premise—admittedly an imperfect one—that shared mythologies and etymologies can reveal deep social memories and shared habits of associative thinking linking certain forms of transport labor to past rebellious acts. Finally, I point to patterns of historical entanglement between political regimes' experiences of rebellious mobility and processes of transport technologies' becoming defunded, abandoned, or destroyed.

Many of my examples come from the same historical period, the roughly hundred years between the mid-nineteenth century and the mid-twentieth century. This was never exactly by design, and there are many exceptions: the book's numerous narratives often extend back in time to the seventeenth or eighteenth century, or lurch forward into the late twentieth or early twenty-first century. Nonetheless, the discussion has a clear historical emphasis on the period between 1850 and 1950. This emphasis merits some comment. For ruling political regimes to reject transportation technologies because of those technologies' real or perceived associations with subversive patterns of human mobility, two historical conditions have to be in place. One, the regime must be in a position of fearing the population, or some segment of the population, over which it exerts power. Two, a new mode of transportation must appear to the ruling regime to provide a replacement technology that is seemingly more politically palatable. It is at least plausible that the late nineteenth and early twentieth centuries, a period of significant overlap between motorized and nonmotorized systems of transportation, as well as between tracked and off-road modes of mobility, had one or both of these conditions in place to a greater extent than previous or subsequent eras. This in turn may help explain why, when I was undertaking this research, the period from 1850 to 1950 proved to be especially fruitful for my endeavors.

Another, perhaps related, explanation for the book's emphasis on the late nineteenth and early twentieth centuries is that, during the years the research was undertaken (2007 through 2013), primary texts from that century-long period were more available than texts from other periods. For texts dating from before the mid-nineteenth century, archival organization, whether physical or digital, becomes more chaotic; after this period, copyright laws severely diminish the scope of searchable material. This is an unfortunate limitation in our digital libraries as they presently exist. But it would not be possible to make or defend the argument in this study without the existence of such resources. As I show in the chapters that follow, a great many historical writers have commented on the usefulness of canal barges for rebels, of mules for bandits, of elephants for insurgents, of waterfront haunts for saboteurs, and so on—but these insights rarely make it into the titles or subject-headings associated with these writers' lasting, archived works. For this reason, I found searches through large digital libraries (Google Books, Archive.org, and the like) to be necessary to ferret out such narratives, even if that strategy meant that the resulting study exhibits certain historical emphases brought about not by original intent but by the strengths and weaknesses of the digital libraries' searchable collections, as these collections exist today.

1 Mules and Upland Banditry

In his 1961 study of effective techniques of guerrilla warfare—based primarily on his experiences during the Cuban Revolution—Che Guevara praises the carrier pigeon as a method for moving secret messages among rebels.[1] But the transport animal Guevara goes into considerably more depth praising for its guerrilla trafficking abilities is the mule. In Guevara's account, the successful rural revolt depends on the establishment of a network of shadow guerrilla factories, supply depots, and smuggling rings throughout the agrarian backcountry, hidden under cover of forest and beyond the reach of the well-traveled roads. Guevara emphasizes the need for backcountry tinsmiths to fabricate plates and canteens, blacksmiths to make horseshoes, leathersmiths to provide cartridge belts and backpacks. He calls for guerrilla shoe factories, armories, saltworks, and dryers; clothing factories, cigarette factories, camouflage fabric factories, and printing presses fully equipped with mimeograph machines, paper, and ink; concealed medical stations with good surgical equipment; and service batteries to repair weapons and "manufacture certain types of combat arms that the inventiveness of the people will create."[2] The farther from heavily patrolled forest highways these shadow factories are, the better. Thus, Guevara advocates the use of mules and muleteers to work the complex, ever-changing web of supply lines among these numerous, overlapping, covert industrial operations. Guevara calls the mule "one of the most useful animals" for the guerrilla force; he cites the animal's "incredible resistance to fatigue and an ability to walk in the steepest zones," carrying more than 100 kilograms for days on end.[3] For the guerrilla fighter the mule is the "best option in rough country," able to "pass through extremely hilly country impossible for other animals."[4]

The guerrilla "muleteers," Guevara adds, "should understand their animals and take great care of them. In this way it is possible to have regular four-footed armies of unbelievable effectiveness."[5] Guevara further recommends the formation of "special teams" or secret "departments of road construction," assigned to cut small trails for mules through the thickest woods.[6] The smuggling of "crops" (food or drugs) from the agrarian guerrilla zone to the cities and ports could be essential for financing numerous aspects of a rebellion. Thus, supply lines must "radiate out from the guerrilla zones" in the backcountry, "spreading throughout the whole territory, permitting the passage of materials" using, in appropriate areas, "mules or other similar transport animals."[7]

It is striking that, at the same time that Guevara was publishing these observations, the U.S. mule-breeding and mule-trading industry was in steep decline. This decline was precipitated in large part by the increasing prevalence of automobiles and tractors among American farmers mid-century, to say nothing of the increasingly industrial, post-agrarian pace and scale of American agricultural production.[8] Yet the agricultural sector was not the only U.S. market for mules. The U.S. Army had maintained lucrative contracts with mule breeders since the early nineteenth century.[9] Moreover, the reasons for the army's decision to deactivate the mule corps in 1956, driving a massive nail in the coffin of the American domestic mule-breeding industry, are considerably more mysterious than the reason for industrial-scaled agriculture's eager adoption of motorized technology. Farmers used mules to haul items, whether plows, crops, or farming supplies, along set tracks and roads, for which motor vehicles (provided an abundant supply of cheap gasoline) seemed better suited, whereas the army used mules in war theaters to transport supplies, or war materiel, across rough terrain, usually far from any fixed tracks or roadways. In World War II, for instance, the army mule corps was pivotal in the Mediterranean and China-Burma-India (CBI) theaters of combat, aiding in the movement of troops and materiel across the dense, roadless jungles and steep mountain environments.[10] In fact, Allied planners of these campaigns severely underestimated the number of mules that would be required in the field. The army had to purchase thousands of new mules from breeders each year in Missouri and Tennessee for export to the combat theaters, buying 10,000 mules in 1943 alone, and often imploring jack stock owners to breed every available jack to a mare.[11] Experienced muleskinners and animal packers were also in short supply and high demand. "Too few men knew how to handle the mules," recalled one CBI theater commander.[12] Another commander remarked that he "would rather have mule packers than infantrymen, for it was easier to make packers into infantrymen than infantrymen into packers."[13] Indeed, as one historian of the American mule has noted, "you can't overnight train saddlers and blacksmiths and pack masters and people of that kind."[14] Knowledge of how to set a saddle right at a moment's notice, how to lace and tie ropes so as to distribute the load evenly over the mule's back, how to detect the need for rest, or whether the mule sensed danger approaching, how to find forage in harsh terrain, how to repair leather tears or apply immediate veterinary aid—such knowledge was often embedded in and dependent on a lifetime's worth of direct personal experience handling mules.[15] Throughout much of the nineteenth century, most working Americans would have some had at least some such experience. By the mid-twentieth century, hands-on familiarity with the art of pack transport was primarily the province of small-time farmers and sharecroppers in the South, mining workers and farmers in Appalachia, mining workers in the West, and finally the army mule corps itself, whose breeding programs and quartermaster offices dated back to the early nineteenth century.[16]

Army spokesmen in 1956 justified the decision to deactivate the mule corps by pointing to recent improvements in tank, jeep, and especially helicopter technology. Helicopter units, read one army press release, would "be superior to similar units using ground transport means" for the purpose of conducting "special operations in the mountains, the Arctic, and the jungle."[17] Many army officials also pointed to the success of the German panzer tanks during World War II as proof that the U.S. military had to become more fully mechanized. Such officials did not remark on the fact that transport animals—mules, horses, and camels—had done the brunt of the work supplying the Germans' far-flung tank divisions with fuel.[18]

Throughout the 1950s and into the early 1960s, any systematic comparison of the actual mobility of mules with the actual or demonstrably achievable mobility of jeeps, tanks, and helicopters came not from proponents of mule deactivation but rather from the mule corps' small but vocal group of defenders and advocates.[19] "Pack animals can go where machines cannot," submitted one Fort Bragg study, exemplifying such advocates' line of argument:

They are by no means outmoded or old-fashioned. Animals have been used for centuries, and in this day of air transportation, are still in use throughout the world. They will be in use long after we establish stations on the moon. Pack animals ... follow the combatant, lightening his load, to the very edge of hell and back.[20]

Some other military proponents of pack transport capability argued that pack mules and motor technology, far from being redundant, could complement each other in theaters of combat. Such proponents pointed, for instance, to field demonstrations in the 1940s that properly trained mules could be safely dropped by parachute into open terrain and that the animals would need only a short amount of recovery time before being able to carry heavy loads. Others pointed out that mule teams could carry disassembled heavy weaponry, such as wheeled howitzer cannons, across untracked areas.[21] But such advocacy merely fell on the "deaf ears of armchair leadership," who, as mule historian Melvin Bradley has put it, "celebrated their accomplishment in ridding the Armed Services of its 'archaic units.'"[22]

The decision by the U.S. Army to deactivate its pack transport capabilities during the 1950s appears especially odd in light of contemporaneous observations regarding the usefulness of the animal not only for guerrilla operations (such as those that Guevara was involved in) but also in counter-guerrilla operations in civil war–torn countries such as Greece and Korea. In Greece, the logistics of the Communist guerrilla campaign during the 1945–1949 civil war were vitally dependent on mule transport for moving goods in secret through the Pindus range, which was the center of the Communist-backed Greek Democratic Army's organizational operations and popular support. The lack of roads in these uplands restricted almost all movement to "foot or mule," and one historian of the conflict estimates that at any given time, as many as 1,400 mules were involved in the off-road movement of Communist supplies.[23] Counter-guerrilla operations in this conflict, conducted by the Western-backed National Army, as well as by American and British troops deployed in Greece to help defeat the Communist insurgency, were also dependent on mules. The mule historian Emmett Essin estimates that the U.S. military supplied 2,500 American mules to the Greek anti-Communist campaign. As one American colonel in Greece, J. C. Murray, later recalled, the ground transport logistics of the American "anti-banditry" war mostly consisted of trucking supplies to the various roadheads at the foot of the Pindus range, and then transferring loads to pack mule teams for haulage up to friendly camps in the fastnesses of the mountains. This was often the only way to outflank the Communist fighters' own mule-borne supply trains through the upland passes.[24]

American troops fighting guerrillas in the forested highlands of the Korean Peninsula were no less dependent on mule transportation. But, as during World War II, U.S. soldiers found themselves undersupplied with mules—in fact, this time no mules at all were sent. Many soldiers resorted to using Korean mules to move barbed wire coils, steel stakes, mines, and ammunition across the hilly terrain. Troops were reluctant to provide commanders with information about how many animals they had with them for fear they would be denied their use.[25]

The final deactivation of the army mule corps occurred shortly after the close of the Korean War, but Korea would not be the last American conflict in which soldiers and field strategists would complain of an absence of adequate pack transport for U.S. troops. During the Vietnam War, some military voices pointed to an overreliance on bombing raids and air surveillance and an underreliance on the capabilities of pack teams working on the ground as an important factor in many American tactical failures in the dense, jungle-covered highlands of Vietnam. One U.S. Army colonel responsible for supplying troops in the jungle argued that pack transportation would have been effective "if it only had been given a chance to prove itself."[26] During the 1980s, some U.S. military commanders, anticipating possible special teams operations in Central America and the Middle East, began to explore the possibility of reinstituting a well-trained pack transport corps within the army or Marine Corp, but, as one writer noted at the time, in the absence of "qualified mule skinners … there's no one in the army today who any longer knows 'which end to hook the rope to.'"[27] The army's 1985 Mules Committee ordered a search for old training manuals to determine what had been "the doctrine on mules" before the 1956 deactivation decision, but deeming the prospect of having to reinvent the mule "wheel" too expensive, the military abandoned the reactivation idea the next year.[28] Indeed, it is only very recently that the U.S. military, citing logistical difficulties with armored Humvees and helicopters in mountainous regions, has once again started tentatively experimenting with training pack soldiers and purchasing mules, this time for deployment in the mountains of Afghanistan.[29]

To answer the question of why, despite pack transport's evident unique advantages for the purposes of off-road cargo movement under belligerent conditions, military commanders elected to deactivate the army mule corps in 1956, a technological "zero-sum game" explanatory model, of the type often emphasized by many economic and technological historians, in which a powerful techno-social "cluster" hinders an institution's capacity to commit to two very different kinds of technologies simultaneously, has some clear, if also somewhat limited, value and relevance here.[30] The mechanization of the military, especially during the 1950s, functioned not just as field-strategic but also as national economic policy, with the secretary of defense from 1953 to 1957, Charles E. Wilson, himself being a former General Motors executive and the relationship between the American military and the large automotive, appliance, and steel conglomerates increasingly resembling what Dwight Eisenhower, in his office-leaving address at the end of his presidency, would describe as a "military industrial complex."[31]

Yet this sort of explanation in and of itself is not sufficient for under-standing either the deactivation of the U.S. Army mule corps or the mar-ginalization of mule breeding from the postwar American economy. Mule breeding and trading were, after all, an industry in their own right. As late as the 1940s, breeders proliferated throughout Missouri, Texas, Georgia, Ten-nessee, and Kentucky. Breeding and trading usually occurred via distinct but overlapping geographic circuits. In one circuit, farmers with mares and jacks of their own would sell mule foals to graziers, who would raise the foals before selling them to traveling traders, who in turn would usually head to markets in the Louisiana Delta region. In another circuit, especially pronounced in Texas and Missouri, mule-feeders owning jack stock would rotate their jacks from station to station within a given town, engaging the use of mares in each neighborhood. In the Kentucky Bluegrass Basin, mule production thrived by taking unwanted weanlings from nearby Ohio—where farmers' "contempt for mules [was] as profound as it was unreasoning"—and "fatten-ing" the animals at Kentuckian barns for eventual sale. Finally, by the 1930s, much mule breeding and trading occurred at the giant consolidated mule terminal markets in Saint Louis and Kansas City, where prized jacks, mares, and mules were transported by boat or train for sale before large crowds of buyers.[32] Before the deactivation of the mule corps, the army acquired many of its mules from these large urban markets, though it also frequently sent its quartermaster agents to many smaller towns and "mule barns" throughout the South and southern plains, keeping track of the complex "mule grape-vine" in each county and region so as to secure contracts for the sturdiest animals for use at the army's numerous depots.[33]

In the late 1940s and early 1950s, though, army agents mostly stopped visiting the market terminals and mule barns. Many American mule breed-ers found during this period that, as they lost their major military contracts, traders from Mexico and Brazil were eager to swoop in on American mule markets and bargain for discounted deals on sometimes tens of thousands of mules at a time.[34] Evidently, during this period, the architects of Ameri-can military-industrial economic cooperation did not register this eager overseas consumer market for American mules as a potential economic opportunity, worthy of institutional support and development. The use of military contracts to place domestic industries in a prime position to cor-ner overseas markets was a common enough theme of American political economy during the 1950s—yet this economic strategy was never applied to the *mule* industry as a way to corner markets in third world countries, where demand for work animals was high and the breeding infrastructure was less sophisticated.

In fact, military leaders' unwillingness to support the American mule industry may have had less to do with economic realities and arrangements during the post–World War II period and more to do with a set of administrative desires and biases predating World War II by at least several decades. As early as 1912, industrial organs like *Motor Age* were already writing approvingly of the War Department's tests "to determine the introduction of the motor truck and the abolition of the army mule."[35] Other papers of this period wrote of the War Department's efforts to "banish" the muleskinners of the Southwest cavalry and "do away with the mules."[36] The U.S. Army's Pancho Villa Expedition of 1916, ostensibly a mission to capture the Mexican revolutionary leader hiding in the mountains of northern Mexico, was purposefully undersupplied with mule trains, as military planners wished to showcase the viability of a motorized, de-animalized cavalry (the expedition was a failure).[37] Despite mules' widespread use by all sides during World War I, at the close of the war the U.S. Army leadership, believing the future of warfare lay in mechanization, sold off the bulk of the army's military-grade mules. The leadership "had forgotten," Emmett Essin writes, "that despite all the modern advances during the first decade of the century, the mule rather than the gasoline engine still dominated the supply lines."[38]

Also traceable to the 1910s–1920s period is a wider cultural bias, or social sentiment, against mules and mule-based transportation. One of the most striking expressions of this bias is to be found on a pair of maps from 1920, showing the prevalence of horses as opposed to mules throughout the United States.[39] As the juxtaposition shows, horses were heavily favored in northern states, while mules were favored in the South. As multiple historians of equine transport in the United States have noted, there is no easy, single-factor explanation for this dramatic geographic disparity.[40] Mules could outwork horses in the more punishing, year-round climate of the South; this certainly helps to explain the hybrid's widespread use there. But mules, by virtue of the superior durability of their back and leg muscles, could also outperform horses in the colder North. Indeed, horses' only main working advantage over mules was speed over long distances—an advantage that was largely negated following the proliferation of railroads and trolley lines in the North during the mid- and late nineteenth century.[41]

Melvin Bradley has described this peculiar regional divide in equine preference as a "puzzle with no answer."[42] Another historian chalks up the difference to some kind of hazily defined "cultural stubbornness" bound up in northern versus southern identity.[43] And to be sure, the historical

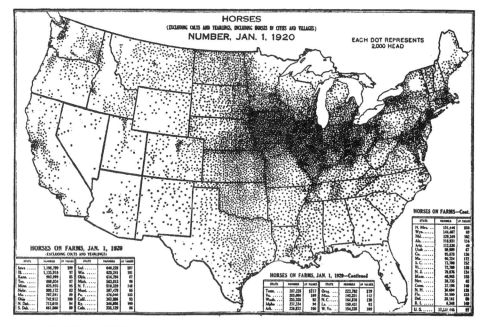

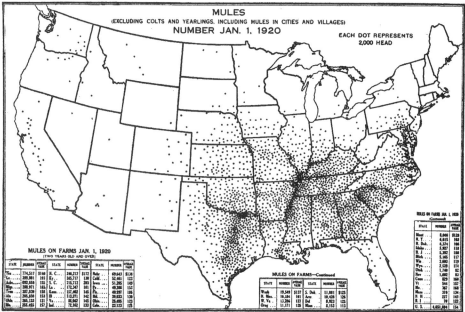

Figure 1.1

The "puzzle with no answer": a 1920 survey showing that horses were heavily favored in the northern states of the United States and mules in the South.

Source: U.S. Department of Agriculture, *Yearbook 1921* (Washington, DC: Government Printing Office, 1922), 472–473.

mule-horse divide in the United States may very well have something to do with a national legacy of political and cultural antagonism between North and South. Yet what is curious is that this divide also extends to Europe. Whereas in northern European countries mules have never been especially popular as compared with horses, in southern European countries, such as Greece, Italy, Spain, and Portugal, mules historically have been overwhelmingly the preferred means of overland animal conveyance.[44]

Animal historians puzzling over the northern aversion to mules have mostly focused their attention on the working abilities of the equine animals themselves, and not on the different degrees of human labor mobilization required for the breeding and trading of mules, as opposed to the breeding and trading of horses. Mules, of course, require biological input from two different kinds of animals: a male donkey, or jack, and a female horse, or mare. This crossbreeding typically results in an animal combining the best qualities from both animals while minimizing the worst. Such crossbreeding is also much more practical when a group of mobile humans is periodically at work keeping the jack stock in geographic circulation among so many stationary mares. The *political* result is that within mule breeding, the mobile traders are likely to wind up with at least as much leverage as the stationary landholders, whereas within horse breeding, the most empowered actors are likely to be proprietors of large manors, who can accumulate both mares and stallions and exchange these animals exclusively among each other. This crucial *human* difference between the geography of mule breeding and the geography of horse breeding may help explain why, in climates where the mule's labor can be passably, if not fully, replaced by the labor of the more fragile horse, landowners steeped in a cultural tradition of landed gentility (and perhaps also a religious tradition traceable to the commandment in Leviticus 19:19: "thou shalt not let thy cattle gender with diverse kind") have eagerly embraced the horse and marginalized the mule, whereas in climates workable exclusively by mules, different social geographies have emerged.

To put this another way, the "contempt" of those farmers in Ohio may not have been for mules directly; rather, it may have been a contempt for the mobile social groups that were associated with mule trading and mule breeding. Chief among these groups were "black gypsies"—American Roma who rotated by the tens of thousands throughout the South and Texas, and were often described as having the power to "hoodoo" the animals.[45] Also closely associated with mule trading were Irish Travelers or "Irish gypsies" (or "traders" or "tinkers"), who often traveled in clans, carrying tents and supplies with them by car, setting up mule markets throughout the South and converging each year by the thousands in Atlanta and Nashville for a common funeral day.[46] Like the Roma of Europe or the inland-canal boating families of the British Isles, such groups had a reputation for keeping to and marrying primarily among themselves. Another group especially associated with the movement of mules was the Melungeons of central Appalachia, thought to be of mixed African, European, and Native American descent (some point to Portuguese and Moorish ancestry), and who had a reputation during the nineteenth and early twentieth centuries for possessing special skills in long-distance herding and trading of mules, horses, and cattle.[47]

ᶜ   If mule breeders and traders have at times elicited associations with wandering clans and racial mixtry and otherness, mule drivers have more often elicited associations with smuggling and political subversion. In his autobiographical *Confessions*, published in 1782, Jean-Jacques Rousseau recalls an episode at his cottage in Motiers, Switzerland, where one evening

> two men arrived on foot, each leading a mule loaded with his little baggage, lodging at the inn, taking care of their mules and asking to see me. By the equipage of these muleteers they were taken for smugglers, and the news that smugglers were come to see me was instantly spread. Their manner of addressing me sufficiently showed they were persons of another description; but without being smugglers they might be adventurers, and this doubt kept me for some time on my guard. They soon removed my apprehensions.... These gentlemen, both very amiable, were men of sense, and their manner of traveling, so much to my own taste, and but little like that of French gentlemen, in some measure, gained them my attachment, which an intercourse with them served to improve.[48]

The men's aims in seeking out Rousseau at his mountain cottage are never made clear, but Rousseau later suspects that the nighttime muleteers were there to subtly encourage him to send his writings to a politically free press in Avignon for publication rather than to Holland, where Rousseau normally sent his work so as to avoid censorship.

In the United States, the figure of the muleteer-smuggler has been more often been linked to moonshine. Appalachian moonshiners' "white mule" variety of corn whiskey may have got its name not only from the "kick" it supposedly delivered but also from the common use of mules for facilitating the movement of mash and sugar up the wooded slopes of the Appalachian backcountry and filled bottles back down. Recalled one West Virginia moonshiner:

> Now they's three different kinds of whiskey. They's white mule, they's moonshine and they's white lightning. Now your white mule, that's when you've got your still

way to hell back in the woods. You have to ride your old mule back in there and carry your sugar in and your whiskey out. Now moonshine, that's when you set up at night and make it. White lightning, that's when you come right out in the open like we do right here and make it in the daytime.[49]

Indeed, during the 1920s and 1930s, the transportation geography of the Appalachian moonshiner's backcountry was not altogether different from the backwoods transportation geography espoused three decades later by Che Guevara in *Guerrilla Warfare*. Instead of concealed shoe and leather factories there were concealed liquor factories, or distilleries. When revenue officers started flying planes over the moonshine mountainsides to locate the "stills," moonshiners resorted to camouflaging their facilities.[50] The principal paths to the stills were mule paths, or "workways," as they were called, which some of the more experienced moonshiners would cut in the pattern of misleading mazes through the brush, to confuse revenue authorities.[51] One writer of the 1930s, describing the latest developments on the "Moonshine Front," noted that

the stills are set up in hollows where there is prime water, or under bushes at the edge of creeks, or in limestone caves, sometimes in smokehouses; one was just a cracker can buried in a sand bar.... The liquor goes out on mule-back. One newspaper reports: "The sheriff's force caught two women riding mules which were laden with twelve gallons of moonshine."[52]

An ex-moonshiner, interviewed during the 1970s, similarly recalled the importance of mule transportation to the clandestine operation:

Had seven men working for me. It took two men to take up the whiskey ... to take care of the still. It took two more to look after the fermenters.... Then it took four men to keep it faared up, to cut the wood with a crosscut saw and tote it. You put mules in there to haul it.[53]

Another noted that mules were doubly useful because a mule "can hear anyone a mile off. He'll point those ears forward just like a pointer dog.... A mule is the handiest thing you ever seen to have around a still."[54]

Other qualities sharpen the parallel between backwoods guerrilla transport as described by Guevara and the transport geography of moonshine production in Appalachia. For many moonshiners, the object of the illicit activity was simply profit; but for many others, moonshine whiskey bore additional political weight. Whiskey making, a form of value-added production one can do without much by way of arable land, was an "inalienable right" in many moonshiners' eyes—"whisky and freedom gang thegither," as the poet Robert Burns put it.[55] In the southern West Virginia mining region of the early twentieth century, the making and trafficking of moonshine whiskey often acted as an important social adhesive among miners of disparate ethnic or religious backgrounds. In Mingo County, for instance, the site of a major coal miners' strike in 1920 at Matewan and a subsequent bloody shootout with strikebreaking thugs, Catholic and Protestant miners, who normally regarded each other with some misgiving, often brewed and drank moonshine together.[56] During long coal-mining strikes, whiskey bootlegging could supplement wages, and union-busting Baldwin-Felts detectives had no more luck locating the miners' stills than the revenue officers did.[57] Nonetheless, for the leading organizers at the strikers' forest camps, the presence of moonshiners in their midst could appear a double-edged sword, on the one hand supplementing wages and helping to transcend some ethnic and religious divides, on the other hand tending to promote alcoholism, petty gang rivalries among different moonshiners, and the overall jeopardization of the strikers' cause.[58]

The extent to which the everyday social and economic operations of militant miners' camps, such as the one at Matewan, may have depended on the smuggling of goods like moonshine, and the extent to which this smuggling was itself dependent on mule-path networks and deep local knowledge in the art of mule handling, surely demand deeper study. It is, to say the least, suggestive that, a year after the Matewan uprising, of the thousands of striking miners who converged on Blair Mountain in nearby Logan County—in what ultimately became the largest armed labor conflict in American history—hundreds brought mules and horses with them to transport battle materiel, much of it leftover war supplies from World War I, in which many of the miners had themselves fought.[59] It's equally striking that much of the subsequent fighting between armed pro-union miners, of whom there were some 15,000 overall, and the 2,000-strong private army raised by local coal mine operators lasted five days and occurred deep in the woods, far from any road or rail line. For decades afterward, large caches of World War I–era battle materiel could be found in hollow trees and dug-out holes across the wooded mountainside.[60] For American industrial and political leaders alike, the rebellious force of angry miners amassed at Blair Mountain in 1921 was intimidating enough to inspire one of the few domestic air bombing raids in American history, which was carried out by federally authorized private aircraft, though President Harding, fearing that more weapons were being smuggled to militant miners in northeastern Kentucky, threatened to send the army's own Martin MB-1 bomber planes if the West Virginia miners did not lay down their arms.[61]

During the same period, thousands of miles from the West Virginia coal-fields, American military officials were experiencing further problems with rebels on mule-back. After Mexican revolutionary general Pancho Villa's 1916 raid on Columbus, New Mexico, American general John J. Pershing led a detachment of 12,000 American troops on the so-called Pancho Villa Expedition into Mexican territory, with the aim of capturing the bandit leader. Military planners intended the expedition as a demonstration of a modern, motorized army, sending very few cavalry or packers across the border, and most of the supplies by truck convoys. This decision kept the full might of the expedition's firepower contained to northern Mexico's sparse and primitive road and track network. Pancho Villa and his army, by contrast, had hundreds of mules and horses with them and were easily able to outmaneuver Pershing by keeping to the mountains.[62] Pershing called off the expedition in 1917. Pancho Villa was assassinated several years later in Chihuahua, yet the perception that mules and muleteers were centrally implicated in the rebellious and criminal movement of cargo in the American Southwest continued unabated. An issue of *Cosmopolitan* in 1922 noted that mules were carrying the load for bootleggers and smugglers around the Rio Grande.[63] Newspapers in 1928 described a pack team stealing gold bars, diamonds, and other jewels at banks in New Mexico and West Texas, hiding in the mountains, and then smuggling the goods across the border by mule, only to be intercepted by gangs of outlaws on horseback.[64] Indeed, while the metaphorical use of the term "mule" to refer to the trafficking of reproductive material across racial boundaries (hence the social category "mulatto") dates from as early as the sixteenth century, the modern-day usage of the verb "to mule" more often refers to the trafficking of drugs or arms across *state* boundaries, usually using one's own body as the secret vessel, and can be traced to mid-twentieth-century smugglers' argot at the U.S.-Mexican border.[65]

Discernible, then, in the history of the American use of mule transport is a pattern of entanglement between experiences of subversive mobility and processes of divestment from, or abandonment of, a specific mode of transport perceived as especially useful for subversive mobility. From the Pancho Villa Expedition through the era of the Korean War, the seemingly contrary perceptions of the mule as irreplaceably useful for logistics in modern ground warfare *and* as something totally obsolete and unworthy of continued investment always appear to be bound up with each other. This pattern of cognitive dissonance reappears numerous times in the chapters that follow. I would suggest that the explanation for this perceptual contradiction was political. Investing in mule-based logistics within the U.S. military would have meant investing in a wider social geography of mule-based transportation in the United States—a geography that could be, at times, threatening to the political status quo, owing to the unique possibilities for subversive mobility which this mode of transportation opened up.

For the U.S. military, the institutional banishment of the mule corps during the 1950s was articulated only in the banal language of progress—this despite the uniquely useful role that mules and muleteers had played in multiple theaters of World War II, in the intervention in Greece, and in the war in Korea. But such a notion of transportative "progress" was connected to a more long-standing social and political distaste, both within and outside the military, for the breeders, traders, and drivers of mules: the mobile gypsies and Travelers, the clandestine traffickers of moonshine and other contraband, the off-road rebels in the mountains and desert. Improvements in motorized transport technology appeared to do away with the need for these sorts of less legible social arenas while still, it was hoped, replicating the mule's physical abilities in off-road terrestrial conveyance.

2 Transportation across Intermediate States of Matter

The usefulness of carrier pigeons and mules for subversive transportation depends in part on these animals' ability to go where fixed roads cannot: to the open sky overhead, or to jagged gradients and remote fastnesses in rough, mountainous terrain. Other geophysical zones are also characterized by this quality of frustrating the implementation of fixed networks of tracks and roads: windswept desert sands, fields of ice and snow, muddy or rain-soaked morasses, and basins of open water. It is to several of the effective methods of transport across such "road-resistant" elemental conditions, and to the special usefulness of these types of transportation for the mobility of insurgents, subversive smugglers, and political fugitives, that I turn in this chapter.

Elephants, *Shat Khats*, and Seas of Mud

In his memoirs of walking with Kachin rebels across northern Burma in 1989, the British American writer Shelby Tucker recalls how, early in the march, he and the Kachins are assisted by a muleteer from Yunnan. The insurgent militia, Tucker relates, is winding its way through the mountains dividing Kachin State, Burma, from southwest China. Not unlike the muleteers described by Rousseau at Motiers or by J. C. Murray in northern Greece, the Yunnanese muleteer's transport niche is transmontane border smuggling.[1]

Also with the rebels is a convoy of elephants. In February, the rebel column shifts from the high mountains along the international border to the wetter Kachin Hills. The Chinese muleteer bids the insurgents good-bye and returns homeward. Across the dense ravines and muddy slopes of the hills, the Kachins' baggage is carried not by mules but by convoys of Asian elephants, whose skilled drivers are called "oozies" in Burmese. Tucker recalls his own surprise at observing the superior mobility of the elephants through difficult, steep, off-road routes, called *shat khats* in Jinghpaw (the main Kachin language).[2] He writes of how, on leaving the village of N'rawng Kawng,

we turned off the main path onto a *shat khat* over a mountain so steep and narrow that even mules would have experienced difficulties. But the elephants shuffled steadily up and over it undaunted, investigating the terrain before them with their trunks and dragging their massive hind legs down the descent.[3]

The term *shat khat*, Tucker explains, dates from the days of British and Burmese soldiers' moving war supplies from India across the Southeast Asian massif into China. At one point Tucker translates the term simply as "short cut," but elsewhere Tucker says that for the Kachins, it means "a deviation from an easy gradient."[4] It becomes clear, in Tucker's diary notes, that the rebel convoy uses the *shat khats* not to save time but to avoid traveling along the valley roads, where they risk being intercepted by the regular Burmese military, the Tatmadaw. Tucker writes of the rebel column's route selection near the border with Nagaland, India:

There were two routes to Pinawng Zup [a Lisu settlement near the Indian border]. The easier route led through the Hukawng Valley and up the Tarung Hka, but strong Burma Army garrisons blocked it at Tarung and Namyung. The other route was top secret and, as the column would have to return by it, had to remain so, but it was known to be extremely difficult, and no column as large as ours could prudently attempt it without elephants. So elephants had to be organized.

The rebels' chief, Kachin Independence Organization council member Seng Hpung, proceeds to purchase elephants from the vicinity, as well as food supplies, since the route is mostly uninhabited. Seng Hpung also sends for a local oozy familiar with this terra incognita to act as a guide. Tucker notes that the brother of this local, rebel-friendly oozy had died fighting for the Nagaland Socialist insurgency some years before.[5]

Later in the journey, in the vicinity of Chaukan Pass, which runs between Kachinland and India, Tucker bids farewell to Seng Hpung and the Kachin rebel column and makes his way across the border into Arunachal Pradesh, India, accompanied by several local guides. On the Indian side of the pass he hires a jeep to take him down a backcountry road toward Assam. The jeep takes "more than two days to travel sixteen miles"—every few hundred yards the vehicle has to be unloaded and extricated from the thick mud of the highlands. Progress by motor transport is slightly hastened when the jeep party is joined by a young engineer from Benares, India, who directs the clearing of the road by bulldozer and the hauling of the jeep across the spongy ground. Observing the bulldozers' plodding along through the

Figure 2.1
Kachin rebel column fording the Tawang Hka, northern Burma, 1989.
Source: Shelby Tucker, *Among Insurgents: Walking through Burma* (London: Radcliffe Press, 2000), 194. Courtesy of Shelby Tucker.

mire, Tucker recalls the relative ease with which, earlier in his cross-country journey, the Kachin rebels' elephants had marched through this landscape of mud.[6]

The use of elephants to move people and goods across muddy, riparian, and flood-soaked physical zones has a long history in South and Southeast Asia. Because of the monsoon season and pronounced spring thaw from the Himalaya, water levels can be extremely changeable within Indian and Southeast Asian river and estuary systems. Moreover, on plains and uplands alike, torrential monsoon rains can turn the landscape into a sea of spongy, blackened soil. Waterways are often too rough or strewn with boulders, sand, and debris to be navigable by boat, and many roads lose the solidity required for wheeled vehicles. Trained work elephants are, within limits, capable of moving through such changeable elements, and so provide access to a wider variety of deluged landscapes than many other transport methods allow. English visitors to the Gulf of Cambay (Khambat) in the seventeenth century noted how sands deposited by monsoon rains and flooding could paralyze the activity of ferries across the bay. A bare road on the mud flats was of limited use as well, since the

ebb and flow of tide at Cambay was exceedingly swift, the sea rising in a moment and in less than a quarter of an hour reaching its full height. This was done with such wonderful swiftness that no horse could outrun it. It came so furiously out of the sea that like a great current it overflowed a vast tract of land.[7]

Under such conditions, elephants were the only transport method capable of wading across the water—though even the elephant's "mahout" (the Hindi equivalent of oozy) had to be cautious not to attempt the great Cambay ford during the peak tidal currents, "when the water flows with greater strength and higher than ordinary," and "it carries and washes away both horse and man, and oftentimes with such force that an elephant cannot withstand the same."[8] In such flood-soaked or tidal physical zones, the transport elephant's unique mobility supplies the human mahout with a transformed geographic perspective on the spatial distinction between land and water. Phuket (once known as Junk Ceylon), a landmass off the western coast of Thailand, appears an island to all but the elephant and mahout, for whom it instead appears, in the words of a 1901 geographic account, "really a peninsula as the narrow strait (Pa Prak) is only half a mile across and fordable by elephants at low tide."[9]

Elephants' ability to haul passengers and baggage with them across floodable areas has made them instrumental as seasonal flood-relief vehicles. One 1969 history of Assam notes how, in the late nineteenth and early twentieth centuries, monsoon rains could cut off tea plantations for months at a time, during which period planters were dependent on elephant transportation to receive supplies.[10] Similarly, elephants and mahouts were essential in the rescue of several hundred British, Indian, and Burmese evacuees from northern Burma who were fleeing the advance of the Japanese in 1942. These evacuees were attempting to make their way from Burma into Assam by way of the Chaukan Pass, but, as a result of the torrential monsoon rains, they became stranded at a bend in the Dapha River. In June, several daring evacuees crossed the torrents by forming a human chain, and then marched into Assam to inform the authorities there about the plight of the larger group back at the swollen river. Previous geographic surveys had declared that the Dapha would be unnavigable by watercraft throughout the monsoon, so a military river boat rescue could not be organized. An Anglo-Assamese tea planter by the name of Gyles Mackrell heard of the evacuees' predicament and, being well connected among the mahouts of the region (mostly from the Adi and Khamti hill peoples), he was able to organize an elephant-mounted rescue expedition to the Dapha and bring back eighty-six evacuees, who had been separated from the larger group and were trapped on a mid-river island.[11] Before running a second mission to rescue the remaining evacuees, Mackrell conducted a survey of the Dapha and determined that river boats should be used as well. The river boats got very close to the evacuees' camp, but because of the rocks, currents, and monsoon mudflows, they had to turn back, and the elephant mission completed the remaining rescues. Remarkably, Mackrell was able to take some film footage of the rescue elephants' slowly inching their way across dangerous river rapids, the white-capped currents up to their tusks, their mahouts mounted above, gauging the elephants' reactions to the elements and steering the convoy accordingly.[12]

The Asian elephant's transportative advantages in these sorts of elemental conditions consist not only in the animal's great size, which makes it capable of fording deeper bodies of water than other pack animals (mules, horses, oxen, camels) can manage, but also in the elephant's complex sensory intelligence in its feet and its trunk. Elephants can strategically expand and contract their feet to adjust to the ground consistency on which they tread; this ability makes them more mobile in muck and quicksand than many far lighter species (or human-built wheeled vehicles).[13] When crossing rocky or otherwise uneven or unstable flooded grounds, elephants use their trunks to sense submerged and treacherous crevices, logs, and slippery boulders. In *Among Insurgents*, Tucker describes how, when fording their way up the Tawang Hka (River) in the Kumon Range, the Kachin rebels' elephants would plunge

into the river at the oozies' bidding, carrying men and baggage on their heads and rumps, towing still others clinging to their panniers, never once evincing the least displeasure at all these parasites [that is, the humans].... With their trunks they explored the always-tricky bottom before committing themselves.[14]

Similarly, a tea planter of the Rydak River region of Assam recalls in his 1935 memoirs how "the changing courses, which prohibited the building of permanent bridges, the deepness of the water and its swiftness ... compelled us to use an elephant as the only method of negotiating those parts of the district lying between the river beds."[15] Crossing the rocky, fast-flowing river courses, the planter's elephant would walk with her head at an angle upstream, "moving inch by inch sideways ... the root of her trunk cutting across the water like the prow of a ship," while feeling "the bottom with her trunk for a secure place to plant her feet."[16]

The trunk of the Asian elephant, in addition to possessing this sensory quality, is also extremely powerful and quite capable of hauling large sub-merged logs out of the way. For this reason, in the nineteenth century elephants were often employed in the construction of log dams. An Eng-lish travel writer describes the efforts of elephant and mahout damming an Indian river with red keneer logs in the 1870s:

> These she [the elephant] placed with the greatest care in their exact positions unas-sisted by any one (directed, of course, by her driver). She rolled them gently over with her head, then with one foot, and keeping her trunk on the opposite side of the log, she checked its way whenever its own momentum would have carried it into the stream. Although I thought the work admirably done, she did not seem quite satisfied, for she presently got into the stream, and gave one end of the log an extra push with her head, which completed her task, the two trees lying exactly parallel to each other, close to the edge of either bank.[17]

Trained elephants are equally adept at clearing mid-river logjams, whether to prevent flooding or to hasten the flow of timber downstream to sawmills. In the Myanmar Timber Enterprise, which owns and harvests Bur-ma's supply of teak, when an oozy directs his elephant to clear such a jam, the action is called "aung." As recently as the 1990s, aung was an important component of the transport logistics of the Burmese teak industry.[18]

The very qualities that make Asian elephants useful for mobility across flooded and debris-strewn terrain also make them useful as a means of off-road transport for smugglers. One journalist at Wang Kha, Burma, a back-woods market camp near the Thai border, observed in 1979 how Karen oozies entering the camp on elephant-back would undo their sarongs to reveal cotton belts containing silver bars. These were then smuggled into Thailand in exchange for manufactured goods in short supply in Burma. Using techniques of off-road mobility, and taking advantage of Burma's closed economy, the Karen smugglers controlled a significant amount of the cross-border trade. Other Burmese materials they carried across the hills included poppy from the Golden Triangle, jade, rubies, and timber.[19]

Teak logging was banned in Thailand in the late 1980s as a result of the ecological consequences of overlogging in the upland north. Without the root systems to hold the soil together, slopes became prone to landslides and valleys to calamitous mudflows during the wet season. In the road-less backcountry of northern Thailand, an illegal, clandestine, "shadow" logging industry persisted. Elephants were essential to these illicit log-ging activities. In the early 1990s, Tak and Mae Hong Son Provinces in the northwest saw a large influx of captive elephants from other provinces and neighboring countries. Since heavy equipment could not access the valuable teak forests in these regions, and since the illicit loggers couldn't build logging roads without attracting the attention of the Royal Forest Department and the police authorities, the ability of elephants to access, uproot, and haul teak logs comprised the logistical centerpiece of the covert enterprise. Elephant conservationist Richard Lair describes how in 1988, the Forest Department's raids on the illegal loggers' camps resulted in the confiscation of 153 cattle, 313 bicycles, 778 motor vehicles—and only four elephants. "The ratio of vehicles to elephants," Lair writes, "does not indicate relative numbers used but rather the elephant's superior ability to melt away into the forest."[20]

As with mule transportation, the features of elephant transportation that are useful to smugglers and illicit tradespeople can also prove advantageous to guerrillas wishing to avoid interception by entrenched authorities, occupiers, or invaders. During the Vietnam War, U.S. bomber planes targeted elephants in the Vietnamese and Laotian highlands to stop Viet Cong guerrillas from using the animals for transport beneath the forest canopy. One pilot later explained his method for distinguishing between nonthreatening and "enemy" elephants in Viet Cong territory: he would look for mud caked onto the animals' legs and bellies; this residue supposedly signaled that the animals had recently been carrying cargo along a muddy forest trail.[21] The usefulness of elephant-based transportation for guerrilla mobility is an important theme in Shelby Tucker's memoirs recalling his march with Seng Hpung's Kachin soldiers across northern Burma in the late 1980s. Bertil Lintner's *Land of Jade*, another firsthand account of the Kachin conflict during the 1980s, similarly stresses the logistical importance of the Kachin guerrillas' clandestine elephant convoys for maintaining communications among villages, for smuggling jade, poppy, and timber out across the international border into China, and for moving weapons and supplies back in. Lintner, along with his wife Hseng Noung and their newborn child, rode alongside the Kachin fighters on elephant-back in the winter of 1986–1987. Lintner briefly mentions staying at one of the Kachin Independence Army's secret elephant-training villages, a collection of huts located somewhere in the Hukawng Valley, far from any road (Lintner is purposefully vague as to the precise location) and populated entirely by "Kachin armymen and mahouts busy training newly caught elephants."[22] In Lower Burma during the same period, a Karen resistance group, the Karen National Liberation Army, employed elephant-based logistical teams of their own.[23]

Perhaps the most instructive instance of this sort of guerrilla mobility via elephant is to be found in the evasive marches of the rebel leader Tantia Topi during India's Sepoy Mutiny (or Great Rebellion) of 1857–1859.

Tantia's marches are especially edifying because, unlike the marches of Tucker or Lintner with the Kachin rebels across northern Burma, both of which occurred in winter, Tantia's and his rebels' most impressive marches occurred mostly during the wet season, in the late summer and early fall of 1858. Two accounts of these marches are particularly noteworthy for shining a light on the pronounced advantages of the elephant as a form of evasive, semi-aquatic, subversive mobility during the season of quagmires and torrential rains. One is Frances Isabella Duberly's *Campaigning Experiences in Rajpootana and Central India during the Suppression of the Mutiny*, which documents Duberly's experiences accompanying her husband, a British soldier, during the great revolt. The other is George Malleson's multivolume *History of the Indian Mutiny*. Malleson was a British officer stationed at Mysore during the mutiny; he subsequently wrote a history of the uprising constructed from numerous soldiers' diaries and recollections.

Malleson speaks of the "black and spongy soil" of central India during the summer of 1858.[24] In late June, Malleson tells us, Tantia kept "shifting course to frustrate his pursuers," the British and Anglo-friendly Indian military forces desperate to stem the growth of what was looking increasingly like a subcontinent-wide rebellion against British political domination in India. Tantia's forces approached the Chambal River, a Rajasthani upper tributary of the Ganges. "Flight and pursuit were alike retarded by the rains," Malleson relates, "which fell during this month with remarkable force, so much so that the river Chambal, swollen to a torrent, barred Tantia's passage from Indragarh to the eastward."[25]

The Duberly account of this scene of action begins with a verse from James Thomson's 1727 poem "The Seasons":

Down comes the deluge of sonorous hail,
Or prone-descending rain.[26]

Duberly writes that though Tantia's rebels were trapped at the Chambal, the British garrison from Gwalior could not intercept them there because the garrison was "stuck fast in the mud."[27] Those on the British side who attempted the march found themselves merely "floundering along a road always up to our horses' knees, and many times up to their girth in black mud." The battalion traversed the mire anyway, but found itself blocked by a flooded nullah (a riverbed that fills only during the monsoon) and forced to wait there without sufficient food supplies. For a day the rain subsided, and these British soldiers were finally able to march on, "through about two miles of deep mud," to a village to rest. However, they camped near a river, the Banas, "not more than a small brook when we first arrived," which became, "when it rained, such a torrent, and ran with such violence, that it resembled a very heavy sea running," spilling into the soldiers' camp.[28] Finally, in August, the rains started to weaken in force, and the British troops, finding the roads more passable, began to converge on the rebels.[29] However, the British commanders were "met by conflicting news" regarding the fugitives' ability to finally flee across the Chambal. As Malleson tells it, the British commanders were "assured by experts that it was absolutely impossible that Tantia could cross the Chambal at that season of the year, and that he was bent on pushing southward." Other agents received word that Tantia was, to the contrary, quite "determined to cross the river." Listening to the "experts," the British commanders decided to direct their troops southward, with the expectation that Tantia would have to move in that direction overland. Instead, Tantia crossed the river, his fighters on horseback and his baggage train on elephants. By luck or by tactical prowess, Tantia had timed his action perfectly to keep the British trapped on the western bank. On learning of their miscalculation, the British garrison "hurried after [Tantia], reached the river after a hard march, only to find it just fordable, but rising rapidly.... Tantia, freed by the rising of the Chambal from all chance of pursuit, halted five days at Jhalra Patan [in Rajasthan]."[30]

Equally dramatic was Tantia's escape across the Betwa River, further to the east, in October 1858. In this season the soil still resembled, according to Malleson, a "sea of black mud."[31] Similarly, Duberly describes the "odorous and slushy mud" of the Betwa region (in Malwa) in early fall, the roads reduced to a mere "wreck of carts" by the heavy and frequent rains.[32] The monsoon by this time was coming to a close, but, according to Duberly, the remaining moisture and autumn heat left so much dew on the tall jungle grass that it was necessary to "wade" through grass and forest "like passing through a river."[33]

On numerous occasions in late September, Tantia's British pursuers expected to have Tantia ensnared, only to find that the rebel army had outmaneuvered them and disappeared into the forest, its only trace a few elephant footprints and gunwheel tracks. Tantia began zigzagging back and forth across the upper Betwa.[34] Duberly recalls that around this time she was sorry not to have the British elephant train, which had been called to another theater of the revolt.[35] Duberly also recalls that, near the Betwa, the British soldiers became especially demoralized, feeling that "the hearts of all are with the rebels." Tantia and some 12,000 fighters were at a town called Chanderi, near the west bank of the Betwa, restocking supplies. In early October the various British garrisons in the area attempted to coordinate an action to cordon off Tantia's forces inside the town. One lieutenant colonel was to move his force from Guna "so as to keep the Western road closed." A brigadier "would have defended the Northern and Northwestern Roads" into the town, and the main general in the area was to move his troops to Chanderi's Southern Road, to prevent Tantia and his fighters from fleeing in that direction. "The only uncertainty ... was the falling of the Betwa"— which the British commanders, repeating their error from the Chambal theater in August, presumed to be "at that time so swollen as to be absolutely unfordable," "an insuperable barrier to the escape of the rebels eastward."[36] The British force from the south was delayed by inadequate food supplies, giving Tantia's army time to watch the water level in the Betwa gradually go down. By the time the British finally arrived in the vicinity of Chanderi, Tantia and several thousand of his followers "had escaped over the Betwa—having built boats for the purpose," while their supplies (which Duberly calls their "treasure and women") "were conveyed across on elephants." Tantia continued on in the direction of Teary (Tikamgarh) in the east.[37] He would not be captured until well into the spring of 1859, near the end of the revolt.

Today, the sites of Tantia's escapes over the Chambal and Betwa Rivers are submerged beneath reservoirs. Hydraulically stabilized, the Chambal and Betwa no longer offer the same tactical advantages to rebels on elephant-back as they did in Tantia's time.[38] This modern outcome stands in contrast to the Tawang Hka and other northern Burmese rivers through which Tucker and the Kachin rebels marched in 1989. These rivers were at that point still undammed, "wild" courses, as they remain, in large part, in the early twenty-first century.

The Rapti River, which is subject to especially high degrees of seasonal water flow variation (and which is becoming hydraulically regulated only in the twenty-first century) was the site of another dramatic insurgent escape on elephant-back during the Sepoy Mutiny, that of rebel leader Nana Sahib, whose convoy fled into Nepal via a daring river crossing in 1858.[39] William Howitt in his 1864 account of this northern theater of the Indian rebellion describes how Nana, on hearing that British troops were nearing him and his force, then dashed, "with elephants bearing himself and his treasure ... over the Raptee into the Terai."[40] By the time the British cavalry arrived, Nana was safely on the other side. Charles Ball, in his *History of the Indian Mutiny*, published shortly after the end of the conflict, describes the condition of the river as the British approached:

The Raptee, then at its lowest, was a very clear, rapid, mountain river, with low banks, between which were beds of sand deposited by the torrents, which had descended from the hills during the rains. The course of the river is exceedingly tortuous, and little or nothing was known of its direction or of the fords.

As the British forces hesitantly made their way across, some of Nana's fighters who had remained behind ambushed them in the water. Ball describes how the sides "struggled in the whirling stream." He adds that "the river was the most formidable foe." After the fighting was done, the British found bodies—both theirs and Nana's—submerged in quicksand downstream from the treacherous ford.[41]

Nana Sahib was never caught. The rebel leader's evasiveness would provide a principal story element for Jules Verne's 1881 fictional novel *The Steam House*, which had the alternative title of *The End of Nana Sahib*. In this story, a group of British Raj soldiers, colonists, and adventurers continues the search for Nana Sahib well after the rebellion's close. The pursuers travel through hostile and subject lands alike in what Verne describes as a "moving citadel," which the main characters affectionately nickname their "Behemoth": a steam-powered, mechanical elephant.[42]

According to Charles Ball's account, shortly after Nana Sahib's escape into Nepal, several lower-ranking rebel chiefs recrossed the Rapti ford on their elephants to surrender. Noted among their ranks was "General Wheeler's sporting elephant, and Mahout, who had been going about with the rebels ever since the Cawnpore massacre." This massacre, the worst the British suffered during the Sepoy Munity, had occurred in 1857. Nana's forces had laid siege to the British garrison town of Cawnpore (Kanpur), on the banks of the Ganges, for several weeks. General Hugh Wheeler, in charge of the British forces in the town, surrendered on June 23. Nana Sahib agreed to let the garrison and their families—some nine hundred people in all—exit the town and proceed to a ghat on the river bank, from where they were to be taken by boat to a British-controlled area. Lieutenant Mowbray Thomson, one of the few on the British side to survive the Cawnpore incident, later wrote of the "mournful procession" of the colonists, conveyed by elephant and palanquin to the ghat. Thomson recalled that to many in the garrison, the whole agreement seemed like a trick: on the approach to the ghat, the mahouts seemed to be conspiring with the Ganges boatmen. When the British were in the rivercraft, a signal from the rebel sepoys seemed to be given, at which point the native boatmen tossed "burning charcoal in the thatch of most of the boats" and jumped overboard.[43] The British soldiers, still armed, began to shoot into the water and onto the shore. Nana's soldiers fired back, and the British, vulnerable in the middle of the river, were nearly all killed.[44]

Associations between elephants and watercraft are observable in other moments of the Sepoy Mutiny. Tantia Topi used both elephants and boatmen in his escape over the Betwa. Continuing into the Yamuna valley in the fall of 1858, Tantia was aided by the insurgent river pirate Rup Singh, whose waterborne depredations on the Yamuna drew some British pursuers away from Tantia's convoy.[45] Elephants and watercraft are similar in their usefulness for rebel mobility. Both are capable of moving stealthily and "off-track," though each under a different set of physical limitations. The rebel watercraft is constrained by the geography of shoals, rocks, mudflows, and logjams; moreover, narrow, straightened, and heavily populated waterways may prove too easy for authorities to patrol. The rebel mahout has curbs on his or her clandestine mobility as well. Elephants leave temporary tracks. Powerful water currents, such as the peak tidal current at Cambay, can cause fording elephants to lose their footing and drown. And, of course, elephants cannot wade across bodies of water more than five or six feet deep. As far as this last limitation is concerned, though, it's worth noting that according to some commentaries, elephants can be used as floating freight vessels, at least on short crossings over relatively calm waters. Georg Schweinfurth, the late nineteenth-century German naturalist, describes how in 1881, six baggage elephants imported from India were marched from Cairo to Khartoum by British colonial officials, "swimming the Nile six times" during the journey.[46] Shelby Tucker talks about the "boat-like" quality of the Kachins' elephants. In Tucker's account, while the rebels are following the course of the Tawang Hka in Kachinland, they find that in places the river is like a "raging torrent," too deep to allow the upstream ford to continue.[47] For a while the rebel column is able to find some abandoned gold-panners' paths and animal tracks to follow near the bank, but eventually the paths disappear into the brush.[48] "Whenever need arose," writes Tucker, the elephants "were superb swimmers." At one bend, when the rebels fail to find a *shat khat* on the east bank, they simply swim their supplies across by elephant alongside a gorge, and find a route up the west bank instead.[49]

Duberly stresses the similarity between waterborne and elephant-borne transportation as well, likening the way a baggage elephant "ploughed through the water" to the "rush of water under the stern of a ship."[50] In Hindu and Buddhist mythology in South and Southeast Asia, elephants are often associated with flowing water, as well as with rain and clouds. In one myth the elephant is a rain cloud condemned to walk the earth. Hope Werness writes, in the *Encyclopedia of Animal Symbolism in Art*, of another myth:

In India, the elephant's watery association stems from its birth—Brahma, the demiurge creator sang over two half egg-shells and Airavata, the divine elephant, came forth.... Airavata was the son of Iravati; Irrawaddy is the principal river of Burma, and Ira means water.[51]

Many British commentators from the period following the Sepoy Mutiny seem also to have grasped that during the South Asian season of torrential rains and mudflows, hard distinctions between land and water, and for that matter between pack animal and watercraft, may become less applicable. Malleson writes of the Malwa region's "sea of black mud" and miles of black "saturated cotton soil."[52] Duberly talks about the river Banas resembling a "heavy sea running" and the dew in the jungle grass being so abundant that even walking through the fields and forests felt like wading through some endless body of water. To convey her own experience of this half-terrestrial, half-aquatic environment in mid-autumn, Duberly quotes a passage from Tennyson:

Thro' scudding drifts the rainy Hyades
Vex'd the dim sea.[53]

Camels and Granules

Geographic descriptions like these—the sea as land, the land as sea—run parallel with an early twentieth-century trope, at least among English writers, of sandy deserts as "sealike," especially to the extent that such deserts could facilitate subversive or piratical patterns of mobility. In her 1916 essay "Pirate Coasts of the Mediterranean Sea," the American geographer Ellen Semple claims that the Barbary pirates of the Moroccan coast hailed from the desert raiders (that is, the Tuaregs) of the inner Sahara. These raiders, Semple tells us,

adapt themselves to the seaboard environment, blend with the local inhabitants, from whom they learn the art of navigation, and pursue their ancestral trade, exchanging the desert camel ... for the swift-moving ship.[54]

Attempting, in the tradition of turn-of-the-century environmental determinism, to establish the "physiographic" origins of human behaviors, Semple links piracy to the sandy desert, where camels are "ships" plying the wavelike dunes and the desert towns are the sand-sea's "ports" in which the camels are loaded, traded, and raided.

Like Semple, T. E. Lawrence speaks, in his memoirs of the 1916–1917 Arab Revolt, of the Arabian desert as a "pathless sea of sand," an "untrodden wilderness" populated by a "gulf stream of desert wanderers."[55] Lawrence describes how the tracks left by the camel-mounted Arab rebels were erased by the elements:

There were no footmarks on the ground, for each wind swept like a great brush over the sand surface, stippling the traces of the last travelers till the surface was again a pattern of innumerable tiny virgin waves. Only the dried camel droppings, which were lighter than the sand and rounded like walnuts, escaped over its ripples. They rolled about, to be heaped in corners by the skirling winds.[56]

Lawrence wrote the "Guerrilla" entries in the 1929 and 1932 editions of the *Encyclopaedia Britannica*. In these he refers to the similarities between "the Desert and the Sea" from the standpoint of the Arab guerrilla committed to maintaining a position of evasive mobility relative to the better-armed Ottoman forces:

In character these operations were like naval warfare, in their mobility, their ubiquity, their independence of bases and communications, in their ignoring of ground features, of strategic areas, of fixed directions, of fixed points. "He who commands the sea is at great liberty, and may take as much or as little of the war as he will": he who commands the desert is equally fortunate. Camel raiding-parties, self-contained like ships, could cruise securely along the enemy's land-frontier, just out of sight of his posts along the edge of cultivation ... with a sure retreat always behind them into an element which the Turks could not enter.[57]

The environmental artist Ned Kahn has spoken of sand as the "granular state of matter"—finely ground-up rock that, being highly subject to wind-based (aeolian) physical processes, does not behave in a very rocklike manner and, if anything, is more comparable to the wave actions of liquid water.[58] Indeed, the similar wavelike behaviors of desert sand and open water are the subject of a classic 1941 study in physical geography, Ralph Bagnold's *The Physics of Blown Sand and Desert Dunes*. During the 1930s Bagnold began to measure the downwind migration of barchan dunes in the desert of western Egypt and northern Sudan. He found that the aeolian processes responsible for these migrations were similar to the physical process of wave formation on lakes or the sea.[59] Moving air, like moving water, is a fluid, though of a less dense variety. Aeolian processes can move quantities of both liquid and solid material, and individual granules of sand can be highly susceptible to such transportation.[60]

In 1980, an American geologist by the name of Vance Haynes found a pile of old tins in the Selima sand sheet of northern Sudan. He confirmed that Bagnold had left the tins there a half century before. Looking at Bagnold's measurements of the local dune topography during the 1930s, Haynes determined that a large barchan dune that had been upwind from the campsite when Bagnold stayed there had spent the intervening half century migrating downwind, covering the tins for several years, then leaving the reexposed detritus in its wake. Haynes calculated that this dune had migrated at a rate of roughly twenty-five feet per year; some dunes migrate at four times this speed.[61] This changeable quality of desert sand dunes presents a special challenge to road engineers in these sorts of physical regions. Unless roads are constantly cleared of encroaching drifts (laying down irrigation pipes and planting vegetation along the sides of the roads can also help), roads in sandy deserts will become engulfed and submerged by the ever-advancing dunes.[62]

In his 1975 study *The Camel and the Wheel*, Richard Bulliet highlights the divergent experiences of the United States and Australia in the importation of camels to aid in the exploration of their respective "sandy desert" frontiers during the mid- and late nineteenth century. The American attempt, Bulliet notes, was short-lived. The first effort at importing camels into the deserts of North America was launched by President Franklin Pierce's secretary of war, Jefferson Davis, during the 1850s. Responding to Davis's urging, in 1855 Congress set aside $30,000 for the development of a U.S. military camel corps. Several camels were indeed imported from the Mediterranean, though the attendant camel handlers taken alongside the camels were not highly skilled.[63] The project was disrupted by the outbreak of the American Civil War, and subsequently its reputation was tarnished through its association with Davis. Yet, as Bulliet remarks, this association in and of itself does not provide a sufficient explanation for the eventual non-adoption of camel transport in the American West.[64] The scheme of importing camels had its northern supporters as well, most prominent among whom was the Vermont senator and ecological writer George Perkins Marsh.[65] Marsh was excited enough by the transportative possibilities offered by camels that he wrote a treatise in 1854 titled *The Camel: His Organization, Habits and Uses*, urging the development of an American camel corps to aid in the exploration and prospecting of the West. However, in this otherwise optimistic volume, Marsh also includes a possible indication as to why, in the nineteenth century, the introduction of the camel to the American desert might have been understood by some as socially problematic. Marsh writes:

The habits of these [American] Indians much resemble those of the nomade Arabs, and the introduction of the camel among them would modify their modes of life as much as the use of the horse has done. For a time, indeed, the possession of this animal would only increase their powers of mischief.

Marsh goes on to suggest that, after these initial social disruptions, the camel "might, in the long run, prove the means of raising [the Indians] to that state of semi-civilized life, of which alone their native wastes seem susceptible."[66]

Marsh's association of the introduction of camels to the sandy desert frontier with the potential for "increased powers of mischief" among this frontier's native peoples likely stemmed from his familiarity with the writings and life activities of the Algerian rebel leader Abdelkader (Abd al-Qadir), whom Marsh cites three times in *The Camel: His Organization, Habits, and Uses*.[67] Abdelkader was a camel expert and guerrilla fighter who led a resistance among the Berber hill peoples against the French invasion of Algeria during the 1830s and 1840s. Abdelkader's successes against Louis Philippe's forces were in part a consequence of the Algerian fighters' superior mobility on camels. Camels aided in the movement of gunpowder and arms well beyond the European invaders' reach; camels also enabled the guerrillas to make rapid movements against the French supply lines and disappear back into the desert.[68] A turn-of-the-century writer on guerrilla warfare, Charles Callwell, likens Abdelkader's tactical prowess to that of the "Kirghiz freebooter Kutebar," who led camelmen in a resistance against Russian invaders in the Central Asian steppe later in the nineteenth century.[69] As with the Sepoy mutineers' and the Kachin rebels' use of elephants or the Mexican revolutionaries' use of pack mules, camels were of limited use to guerrillas during combat but very useful in the secret mobilization of supplies, in the outmaneuvering of a better-armed, road-bound enemy, and in the off-road linking together of rebellious cells that were otherwise separated by geographic distance and by patrolled roads.

The lack of interest, in the late nineteenth century, in bringing camels to the American West contrasts with the efforts of British colonists in Australia from the 1860s until the 1890s. The importation of camels to help explore, survey, and settle the Australian interior began in 1860 with an expedition launched by Robert Burke and John Wills, who brought six camels from Arabia. These animals were soon deemed too small for the freight-hauling needs of settlers in the Australian outback, so the next expedition imported two dozen camels from Peshawar, in the Northwest Frontier Province of British India, along with three skilled camel handlers from the same region. At this time the British garrisons in India were expanding their horse cavalries, so ships carrying camels from Karachi to Australia's ports often made the return journey with horses.[70]

After the early 1860s, shipments of camels and camel drivers came mostly from Karachi and Baluchistan rather than from the Northwest Frontier Province, but the camel handlers hired out of these areas to make the ocean voyage were still always called "Afghans" by the British settlers in Australia. According to Bulliet, the typical importation pattern was that one "camel Afghan" was brought in for every three to eight camels.[71] The

Afghans' camel-driving and packing skills were deemed essential to the enterprise of prospecting the Australian interior. During the heyday of the Australian camel transport system, there was a community of several thousand Afghans in the outback.[72]

In 1964, a retired white cameleer named H. M. Barker published his memoirs, titled *Camels in the Outback*, recalling his younger days, during the 1900s and 1910s, working with the Afghans to move bore pipes to wells being dug far from the country's sparse rail and road network. Barker discusses what he understood as the Afghans' special skills in camel packing:

The Afghans stuff their saddles with straw envelopes which, in the outback, encased beer bottles. They were put in through slits in the outer lining and hammered home with a strong wooden driver and a mallet. It was a tradesman's job that every Afghan but no white man seemed to know.... When the white men took on pack camels they seldom succeeded.[73]

Barker also marvels at the Afghan camel trains' capacity for stealthy cross-desert mobility. The "Ghans," Barker relates,

travelled quietly and swiftly. They were fond of taking short cuts, would follow up dry creek beds to avoid hills on the main roads and pass through unseen. Often on hot nights when I was camped with the wagon by the side of the road, a string of pack camels would pass, travelling by night to avoid the heat. In the light of my camp fire I would see the camels' legs as they walked past noiselessly. The only sound was the creaking of the loads they carried and the call of the Afghan leading them. When the last camel had passed, the Afghan in the rear, usually on a horse, would give another call, probably to tell the one in the lead that the roadside camp had been passed and all was well. Then silence fell again, while six or seven tons of loading went towards some country store, pub or station.[74]

Despite these abilities at clandestine, off-road movement across the desert, at no point do the Australian camel Afghans seem to have been perceived as especially implicated in the smuggling of contraband in the interior of Australia.[75] In fact, during the late nineteenth century the main instance of subversive mobility via camel that British colonists had to confront was not in western Australia but on the other side of the Indian Ocean, in Sudan and Egypt, during the 1880s. The activities of the Mahdist guerrilla leader Osman Digna provide one example. In Sudan, Osman mobilized his camel-mounted Mahdist rebels to disrupt British efforts to link Khartoum to the Red Sea port of Suakim by rail. One of Osman's more dramatic victories was at Tofrek, a desert area just outside the port of Suakim, in 1885. Walter Truscott, a writer and illustrator who visited Suakim during the revolt, tells of how Osman and his "guerrilla spearmen and camelmen" seemed to "rise

Figure 2.2
Top: Camel train in the Australian West transporting bore pipes, circa 1900.
Source: H. M. Barker, *Camels and the Outback* (Melbourne: Sir Isaac Pitman and Sons, 1964).
Bottom: Cameleers of the Arab Revolt, 1917.
Source: Imperial War Museum Photo Archives Q 58939.

out of the sands of the mysterious desert" on "the whirlwind that lifted its column of brown dust in fantastic shape."[76] Another writer describes how "like a surging torrent the Dervishes descended upon" the unready British, Anglo-Indian, and Australian regiments stationed in the area.[77] To some extent, such language, likening Osman's Mahdist cameleers to a "surging torrent" or to a "column of brown dust," can be taken as rhetoric aimed at equating the rebels with untamed, hostile forces of nature. At the same time, such tropes also serve to highlight a real physical parallel between the mobility of rebel camel drivers in the desert and the spatial mobility of natural elements such as windswept sands and water. This parallel is augmented by a 1913 writer who suggested that Osman's camel-mounted forces had received their arms from sea pirates, for "the amount of smuggling in firearms that had been going on along the Red Sea was amazing."[78] The transshipment, or handoff spot, between Red Sea trafficker and Mahdist cameleer was likely the "Venice of the Red Sea" (so called by Truscott), Suakim itself.[79]

Just after these depredations in northeastern Africa, political attitudes in Australia regarding the camel Afghans began to shift—a succession calling to mind George Perkins Marsh's sense, conveyed in his writings of the 1850s and based on his knowledge of past activities of Arab rebel cameleers such as Abdelkader, that the presence of camels on the American sandy desert frontier might serve to increase the "powers of mischief" of that frontier's rebellious political elements. The largest shipment of camels and Indian and Baluchi cameleers to Australia occurred in 1893, but by this point business leaders and politicians had begun to agitate for a prohibitive tax against the Afghans, which was soon accomplished, and even for an outright ban against further camel imports.[80] This attitudinal shift regarding the desirability of camel transport in the Australian outback seems to have been in part related to the anti-immigrant feeling sweeping the Australian establishment during the 1890s and onward. While most of the xenophobia was directed against incoming Chinese laborers, some political leaders represented the camel-driving class as the worse social menace. One business leader offered, "To my mind the Afghan is more obnoxious than the Chinaman. The Chinaman does sometimes go on the land and do a bit of work, but the Afghan never does."[81] A member of the Australian Parliament echoed the sentiment, declaring before the chamber that "we wanted the camels, not the Afghan drivers."[82]

In the early twentieth century, the Afghan trade began to fall into decline, before virtually disappearing in the 1930s.[83] Much of the Afghans' business was taken away by the expansion of the country's rail and road networks (though even today, in the twenty-first century, such networks cover only a portion of the continent's vast interior), as well as by the eventual diminishment of easy opportunities for surficial exploration and surveying work. Nonetheless, in his memoirs, the retired cameleer Barker posits that the decline of camel transport in Australia was not inevitable, but rather reflected a lack of political will. He writes that "another fifty years' breeding would have seen a vastly improved type of draught camel" for moving goods across the difficult western Australian terrain, and for encouraging new uses of the outback's immense expanses and resources.[84] But, after the initial period of experimentation and excitement associated with the late nineteenth century, few were willing to embark on, invest in, or agitate for any such new program of development.

The Asian Elephant in Africa: Paths Not Taken

Let us turn to another example of divergent colonial attitudes regarding the desirability of importing Asian pack animals to aid in the exploration, surveying and settlement of a new colonial frontier: the case of the Asian elephant in equatorial Africa. During the late nineteenth-century European land grab in Africa, Belgium's King Leopold II grew eager to establish an elephant-based transport service in the Congo Basin, so as to aid in the development of the region as a more efficient resource prospecting and extraction zone. The interest of Leopold and his officials in elephants was likely sparked by the findings of the Portuguese-backed Pinto, Capelo, and Ivens expedition of 1878, which had explored the Kwango and Kasai tributaries of the Congo, as well as the portage area between the Congo and Zambezi river systems. In these upper reaches of the Congo watershed, the Pinto expedition reported inundated plains in all directions, a vast "paludal morass formed by the floods of the Coango, which submerged the savanna … as far as the eye could see."[85]

Other fording animals, such as horses, mules, oxen, or camels, could not be used in this part of the world because of the prevalence of the tsetse fly. For overland transport, European explorers and settlers in equatorial Africa during this period resorted to the use of human porters—mostly kidnapped indigenous labor. Such a system of transport was deemed objectionable by Leopold's explorers because the porters would always desert the expedition at the first opportunity. The Belgian officials, feeling that "some different method of transportation from the worthless porter-system was demanded," were determined to employ "trained Indian elephants" and to import mahout handlers from southern Asia. African elephants are impervious to the tsetse fly, so it was hoped by the administrators that the same might prove true of the more domesticable, work-friendly Asian elephant. Such an experiment had also been recommended by a French geographic expedition to Loango (Gabon) during the 1870s. Furthermore, the Belgian officials were aware that in 1877, a British colonial administrator in Sudan, Charles George Gordon, had marched six Asian elephants from Cairo to Equatoria, Sudan. These animals had died or were in poor health at the end of the tour, though, because Gordon, wishing to show that transport elephants "do not require Hindoo mahouts to manage them," had refused to import mahout labor from India to aid in the experiment. The Belgians, hoping for a better outcome, were determined to bring in the necessary skilled labor.[86]

⌐ In 1879, Leopold employed an Englishman to bring four Indian elephants, as well as thirteen mahouts, from Ceylon to Zanzibar, and from there to Dar-es-Salaam. From here the party set off "through difficult jungles and morasses" in the direction of Karema on Lake Tanganyika to gauge the viability of an overland elephant route linking the Congo Basin (which starts at the lake) with the Indian Ocean. Concerned about the dangers of the expedition, several of the mahouts turned back midway. After this point, the elephants began to get sick. According to observers at the time, this was to the result of human mismanagement of the animals rather than the tsetse fly.[87] Along the way the expedition established an experiment station at a place near Tabora called Simbo or Simba and asked several of the mahouts to stay there and attempt to train the local African elephants for economic work.[88]

During this early period of experimentation, the British showed some comparable signs of enthusiasm for using elephants for transport in their own African domains. During its invasion of Abyssinia (Ethiopia) in 1868, the British military utilized forty-four Asian elephants, with attendant mahouts hired out of India, for hauling guns and other heavy supplies across the hilly, difficult terrain. The British war office's record of the invasion noted that the elephants "proved most useful" for the expedition, but that the region was also dry and under-vegetated for Asian elephants. The report recommended letting elephants search for their own fodder as much as possible, "if the mahouts can be trusted."[89] In 1879, Britain's Royal Geographical Society suggested that "a few Indian elephants, and their drivers" should be taken from India to East Africa to assist in the construction of a trunk telegraph line "from Khartum to Pretoria" (this line was never built).[90] During this time, some expressed their sense that Indian mahouts ought to be brought into the British East African colonies as well. For instance, in

1881, Emin Pasha, the colonial official in Equatoria (South Sudan) who had received Gordon's six Asian elephants from Cairo, wrote in a letter to a colleague:

You ask after the elephants. There are [now] three of them ... but [they] are scarcely used for transport at present, for want of fit persons to attend to them. It was the greatest mistake that Gordon could have committed, for the sake of a few guineas, to send back the attendants that came with the elephants from India.[91]

After this early period of experimentation, though, during which the necessity of importing skilled mahout labor from India became well established, British attitudes began to shift against the idea of using elephants for off-road transport in Africa. A 1902 correspondence among colonial officials about the idea of using baggage elephants in Uganda and British East Africa (Kenya) cites the imperative of importing Indian mahouts as a reason to forgo the project, and suggests that, for off-road transport to remote areas and the preserves of the forest department, the use of human porter labor would be preferable and would prove more cost-effective.[92] In 1909 the governor of Uganda, Hesketh Bell, attempted a small experiment of his own, importing a single elephant and attendant mahout from East India. Bell describes how, while he was fond of the elephant, he took an immediate "unreasoning dislike" to the mahout, whom the governor viewed as insubordinate. Like his fellow officials in the British East African colonies, Bell ultimately concluded that, in Uganda, the use of native porters would be preferable to a transport system based on elephants and mahouts.[93]

Curiously, during this same period in which British colonial officials in East Africa were dismissing the idea of importing an elephant-driving workforce from India, British merchants, planters, and administrators in Burma were at work organizing the importation of Indian mahouts to work the teak plantations of the Burmese forest interior. Shelby Tucker notes that some of the command words used by the oozies of the Burmese teak forests are of Bengali or Tamil origin.[94] These date from the days of British colonials' importing mahouts from abroad—which they did even though Burma already had in place a long-standing indigenous tradition of elephant capture, training, handling, and driving. The disjuncture is striking: send Indian mahouts to the country where skilled elephant handling labor already exists; caution against sending Indian mahouts to the region where the skilled labor required to train and drive elephants does *not* exist and where the disease-carrying tsetse fly prohibits the use of alternative transport animals. Perhaps in Burma, which the British conquered relatively late, the Indian mahout was understood as in some sense the "lesser of two evils," since by this time the British colonists had at least some ongoing familiarity and working relationship with the culture of Indian mahoutship. By bringing in the Bengali and Tamil mahouts, the British authorities could hope at least to get some handle on elephant transport in Burma, rather than entirely cede this social realm to the more unfamiliar Burmese oozies.

The Belgians never had to endure the experience of seeing their own troops outmaneuvered by hostile mahouts, as had happened to the British during the Sepoy Mutiny. And, in contrast to the British hesitance to establish an elephant transport system in East Africa, the Belgians, following Leopold's initial 1879 experiments, persevered in an "almost unbroken record of attempts at elephant training" in the Congo Basin. Research stations were established near the Dungu River. First Indian and, later, Sri Lankan trainers were hired to train local people in the craft of mahoutship and the language of elephant commands.[95] Unlike the British, the Belgians lacked their own elephant extraction zone in South Asia, so the Belgian elephant enthusiasts (perhaps emboldened by the notion that in classical times Hannibal had used African rather than Asian elephants to attack Rome) soon shifted their energies to the capture and training of local African savanna elephants rather than Asian elephants from abroad. These experiments were very slow-going, not least because elephant generations are roughly twenty years long, severely restricting any potential for a selective breeding program. The same restriction applies in South and Southeast Asia as well, where despite millennia of the human use of elephants, no domestic breed has ever emerged as biologically distinct from the wild species.[96]

Nonetheless, in the Belgian Congo, the elephant experiments progressed well into the twentieth century. A Belgian filmmaker, visiting the central research station at Gangala-na-Bodia in 1935, reported that the local mahouts, now consisting of many Congolese alongside the Sri Lankans, had "developed methods of their own, very different from anything I have ever seen in India."[97] The Gangala-na-Bodia station's heyday was during the 1950s, when eighty-four African elephants did various jobs, including plowing fields, hauling timber, and pulling carts. After this point, capture and training activities were abandoned owing to the numerous civil wars following independence. By 1980 only four elephants remained at Gangala-na-Bodia, but in 1987 Congolese park officials tentatively began to reimplement the elephant training programs dating from the Belgian period.[98] In light of the modest achievements of the Gangala-na-Bodia mahoutship programs, it is an open question how much more successful the British, with their considerable advantages in the Indian Ocean trade system, might have been had they shared the Belgian colonials' ongoing eagerness to implement elephant transportation in Africa.

Many-Headed Monsters and Guerrilla Sled Dogs

In their study of the spatial circulation of radical political activity in the North Atlantic economic realm of the seventeenth, eighteenth, and early nineteenth centuries, the social historians Peter Linebaugh and Marcus Rediker invoke the myth of Hercules and the many-headed hydra. During this period, they write, "the classically educated architects of the Atlantic economy found in Hercules … a symbol of power and order." For imperial rulers, the mythical twelve labors of Hercules symbolized vast economic and geographic ambition: "the clearing of land, the draining of swamps, and the development of agriculture, as well as the domestication of live-stock, the establishment of commerce, and the introduction of technol-ogy." Rulers placed the image of Hercules on "money and seals, in pictures, sculptures, and palaces, and on arches of triumph."[99] Yet, as Linebaugh and Rediker note, these same rulers of the Atlantic economy found in the figure of "the many-headed hydra" a foil for this "Hercules" of imperial power— an "antithetical symbol of disorder and resistance, a powerful threat to the building of state, empire and capitalism." In ancient myth, the destruction of the venomous Hydra of Lerna was the second labor of Hercules. The crea-ture, named for the very quality of fluidity, was "born of Typhon (a tempest or hurricane) and Echidna (half woman, half snake)," and was one of a band of many-headed monsters that included Cerberus and Geryon. For British imperial elites from the beginning of colonial expansion through the early phases of industrialization in the early nineteenth century, the figure of the hydra stood for the specter of social combination and politi-cal organization among an emerging, increasingly self-aware, transatlan-tic working class of dispossessed and otherwise marginalized, downcast groups. Magistrates and officials variously designated "dispossessed com-moners, transported felons, indentured servants, religious radicals, pirates, urban laborers, soldiers, sailors, and African slaves" as the numerous, ever-changing heads of the feared social monster.[100] The hydra myth stood for "the volatile, serpentine tradition of maritime radicalism" that "would appear again and again … slithering quietly belowdecks, across the docks, and onto the shore, biding its time, then rearing its heads unex-pectedly in mutinies, strikes, riots, urban insurrections, slave revolts, and revolutions." Waterborne rebels working on ship and dock engaged in the smuggling of arms, fugitives, and insurrectionary pamphlets, along the dominant currents and fluvial and estuarial arms of the North Atlantic Basin.[101]

The word "hydra," in addition to referring to a monster from Greek myth, also means "water," and it is on historical episodes of waterborne subversive mobility, via merchant vessel, pirate ship, or coastwise boat, along docks and within waterfront haunts, up canals and rivers and into swamps, that Linebaugh and Rediker focus the bulk of their analysis. But these authors' interpretive framework—identifying a historical perception of "hydra"-like, ungovernable patterns of mobility and rebellious geographies of evasion and connection—could as easily apply to the animal-based human mobilities I've looked at so far: that is, to carrier pigeons transporting secret messages across the sky, to muleteers smuggling cargo through jagged, road-defying uplands, to bandit-cameleers crossing sealike desert sand dunes, and to elephants and mahouts evading capture while moving across monsoon-soaked "seas of mud." A Sanskrit word, *naga* (नाग), captures some of the spirit of such an expanded analysis. In some Indian and Southeast Asian mythologies, the *naga* is an entity much resembling the hydra, a many-headed serpent associated with water. The same Sanskrit word can also mean "elephant."[102]

The cases of subversive mobility by transport animal I've considered so far have been from the mid-nineteenth century on. But it is also possible to expand on the physiographic dimensions of Linebaugh and Rediker's hypothesis while remaining within the time frame that interests them, the seventeenth and eighteenth centuries. For instance, one of Linebaugh and Rediker's most prominent examples of waterborne subversive mobility arises in the case of the Great Negro Conspiracy in New York City in 1741. This revolt involved West Africans, Irish, Caribbean Islanders, and Native Americans who had wound up along the New York waterfront by following, or being traded along, the North Atlantic waterways. During the uprising, the rebels combined the experiences of "the deep-sea ship (hydrarchy), the military regiment, the waterfront gang, the religious conventicle, the ethnic tribe or clan to make something new, something unprecedented and powerful."[103] The insurgents smuggled weapons to secret rooms in a waterfront tavern (nicknamed "Oswego" after an Iroquois frontier fur trading post) on the city's west side, plotting the establishment of a "motley" (multiracial and egalitarian) government in New York.[104]

At the same moment that these watery spaces—ship, dock, waterfront tavern, river, and ocean current—were helping to facilitate the coordination of rebellious political energies among different outcast social groups on the Atlantic seaboard, on another frontier of empire, in the Far East of Siberia, the unique spatial mobility of sled dogs on snow and ice (water in its cold, crystalline form) was helping the Chukchi people and their neighbors coordinate an anti-imperial resistance of their own. In the late seventeenth and early eighteenth centuries, the Cossack armies employed by the tsars, having followed the dwindling fur-bearing marten populations eastward, arrived in the remote lands of Kamchatka and Chukotka. The Cossacks moved mostly on horseback, on reindeer sledge, and on foot. The major indigenous groups of the coastal regions—the Nivki, Itelmen, and Koryak of the southern shores, the Chukchi of the Bering and Arctic coasts—were sled dog–breeding peoples who developed their dogs to move long distances over deep, powdery snow and across long stretches of sea ice. The Nivki people bred their dog to pull sleds between their year-round villages on the Amur River and the winter sea-mammal hunting grounds one hundred miles away, on the island of Sakhalin.[105] Similarly, the coastal Chukchi (not to be confused with the inland Chukchi, who were nomadic reindeer herders rather than semi-sedentary dog breeders) developed their dogs to "whip across deep snows" and cover long distances, enabling a single hunter to ice fish on the remotest patches of sea ice and return to his inland village with his catch on a single trip.[106] Not unlike elephants' sensory abilities when traversing uneven river bottoms, sled dogs are often capable of sensing dangerous sea ice fissures that are invisible to the human eye. A skilled dog driver (or "musher") reads the dogs' reactions to the ice and guides the sled accordingly.[107]

In 1697, Vladimir Atlasov was the first Russian commander to experience the indigenous Far Eastern Siberian use of sled dogs for guerrilla warfare. Atlasov set off to subjugate Kamchatka with several hundred Cossack soldiers. The peninsula's indigenous defenders—the Yukagir, Koryak, and Itelmen—descended on Atlasov's forces on dog sleds and skis to aid speedy attacks and rapid retreats, one man firing arrows from a sled while a second drove the dogs.[108] Nonetheless, after two years of such hostilities, Atlasov was able to conquer most of Kamchatka.[109]

Emboldened, and with their eye on the fur trade, the Russian imperial forces began their advance northward into Chukotka in the 1730s, and the Chukchi sled dogs, like their counterparts on the southern coasts, became key operatives in the native resistance. The Chukchi dog breeders, in recurrent communication with the Chukchis' southern neighbors, had spent the decades after the Russian conquest of Kamchatka breeding the Chukchi dog in anticipation of Russian northward aggression. The Chukchi dog was smaller and more energy-efficient than most other northern breeds (though the Koryak dog was comparable).[110] Chukchi sled dogs could scamper up precipices and see at night, which gave the indigenous fighters superior mobility during the long darkness of Arctic winter.[111] The Russian commander Dmitri Pavlutski, or "Yäku'nnin," as he was known among the Chukchi, led his Cossack forces into Chukotka in 1731, with orders to "attack and uproot completely" all Chukchi insurgents.[112] Native resisters were to be killed and peaceful Chukchi deported south to be Christianized at the Russian fortresses. The Chukchi fighters never approached Pavlutski's army in large numbers, instead leaving the Cossacks to wander aimlessly about the tundra fields while the resisters moved strategically by means of dog teams and reindeer.[113] Lorna Demidoff and Michael Jennings explain how,

when pressed, the dog-breeding Chukchi would simply move the population of an entire village out over the ice to hunt seal until the Russians had gone. One village of Arctic coast Chukchi was officially destroyed when its inhabitants abandoned their winter dogfood stores of frozen walrus and disappeared out over the Arctic pack ice. Six months later, just before ice breakup, the entire village returned in better shape than before.[114]

Unhappy with Pavlutski's progress in the Far East, in 1742 the imperial government declared a policy of total destruction of the Chukchi people. All Chukchi men, peaceful or not, were to be killed, and the women and children captured and deported. Once more, when Pavlutski's soldiers approached a Chukchi village, the villagers simply escaped out toward the Bering Strait with their dog teams. At the edge of the sea ice, they moved their dog teams into skin boats and disappeared across the water, likely "to their trading partners, the Alaskan Eskimos." Pavlutski tried again in the summer, hoping that without the ice, the villagers would be less mobile, but the Chukchis simply loaded into their boats again and pushed off into the sea. In 1747 the Chukchi insurgents succeeded in trapping and killing the entirety of Pavlutki's force in a bloody ambush.[115]

The Chukchi sled dog then largely disappears from the written record until the twentieth century. The silence owes in no small part to the success of the Chukchi resistance movement. In 1837 the Russian government, fatigued by repeated failed attempts to subject the Chukchis to direct rule, signed a treaty granting the Chukchis indigenous control over all the coastal regions and most of the interior regions of the Chukchi Peninsula. The only treaty of its kind within the Russian Empire, it exempted the Chukchis from paying taxes and banned Russians from entering the Chukchi region.[116] This treaty likely saved many Chukchis from the ravages of smallpox during the rest of the nineteenth century, as this was the only indigenous group in the Siberian Far East to expand in population size over this period of time.[117] The curtain around Chukotka also prevented Russian writers from visiting, and it is likely for this reason that the Chukchi dog fades from archival records until the turn of the twentieth century. During the first decades of the twentieth century, some dog-mushing enthusiasts from America visited the Chukchi region and, intrigued by the potential of the Chukchi dog as a racing breed, exported specimens to North America.[118]

After the Russian Revolution, the dog-breeding Chukchis lost their independent political status. Starting in the early 1930s, Soviet officials began a campaign to replace the indigenous long-distance sled dog with a freighting breed deemed more acceptable by planners of the Soviet Arctic economy.[119] At this time, most Russian observers perceived the Chukchi dog as the poorest of all the indigenous dog breeds. This was because of the Chukchi dog's small size and inability to haul large quantities of freight over short distances, to railheads or barge landings.[120] Of course, the Chukchi dog's compact size was precisely what had made it such an effective long-distance dog, capable of outmaneuvering horse- and reindeer-mounted Cossack on harsh tundra terrain and open sea ice. When the Chukchis needed to haul heavier loads, they would simply hitch more dogs to their sleds.

Soviet officials wanted a standardized northern breeding program designed around four desirable canine types: freighting dogs, whose specialty would be hauling heavy batches of cargo over short distances; big-game hunting dogs; small-game hunting dogs; and reindeer-herding dogs. Standards for each of the four types were established in Leningrad over a three-year period by averaging the measurements of four hundred representative dogs. The Chukchi dog, considered too small for the freighting requirements, was excluded from the sample. A Soviet dog nursery was opened at Chaun on Chukotka's Arctic coast in 1935 to introduce the Leningrad-approved, pedigreed breeds. Further breeding of the indigenous Chukchi dog was declared illegal.[121] During the bloodiest years of forced collectivization, Chukchi village leaders who refused to cooperate with the new dog laws were frequently killed by police. After World War II, the All-Russian Cynological Congress declared even the "Leningrad factory" sled dog undesirable and campaigned to forbid the future breeding of all sled dog types in the Soviet Arctic.[122]

Pidgin Coalitions

The success of the Chukchi resistance movement of the eighteenth century hinged in large part on the guerrilla mobility enabled by the Chukchis' unique type of sled dog. The Chukchi success was also likely much aided by the Chukchis' historical role as intercultural middlemen in the indigenous trade networks of Far Eastern Siberia and on both sides of the Bering Strait. A pidginized form of the Chukchi language served as a lingua franca in contacts among the Chukchis, the coastal Alaskan Eskimos, the Koryaks, the Kereks, the Yukagirs, the Evens, and the Evenkis.[123] This communicative geography, which was itself likely a partial outcome of the Chukchis' geographic mobility through their specialized long-distance sled dog, helps to explain the Chukchi ability to "disappear" onto the ice and sea, and to enlist the help of their trading neighbors.

Linebaugh and Rediker point to the phenomenon of "pidgin communication" for understanding the ability of dispossessed and conscripted peoples of divergent linguistic backgrounds in the North Atlantic to form coalitions and cooperate with each other in the organization and execution of mutinies, escapes, secret shipments, and violent revolts. In everyday economic life, go-between pidgin tongues functioned both to ease the logistics of global maritime trade and to "commonize" the work of linguistic translation.[124] In the Caribbean, Linebaugh and Rediker observe, news of freedom and revolt, aided by pidgin methods of communication, could spread rapidly among the oppressed, in particular through physical encounters on the waterways. When one Jamaican plantation owner, intent on freeing her slaves in 1817, was urged by the governor to reconsider, she explained that it was already too late:

Word of her intentions had already got around: "I told them not to speak of it, but they talked of it more. The news is gone to Old Arbore and St. Anns, to the Blue Mountains, to North Side, and the plantain boats have carried the news to Port Morant, and Morant Bay."... "The way to go was by water, along the trenches, canals, and rivers, and along the coastline, from one estate's barcadier or jetty to another, in all manner of small craft, manned by slaves who heard and carried news."[125]

The very etymology of the word "pidgin" may capture the specific usefulness of pidgin communication in aiding the formation of coalitions among oppressed groups separated by language and geographic background. Varying etymologies of the term suggest it is a Chinese pronunciation of the English "business," an English pronunciation of the Portugeuse *ocupação*, or an English pronunciation of the Portugeuse *pequeno* (referring to the *pequeno português* spoken by coastal peoples of Angola). Another theory points to the Yayo word *pidian*, meaning "people," as an origin, and another still points to the Hebrew *pidjom*, which means "barter." The linguist Lareto Todd theorizes that it is possible that these meanings and associations "were further reinforced by the English 'pigeon,' especially if trade varieties of English were equated with parrot-like repetitions." Todd points out that the spelling for the bird and the spelling for the language were at times interchangeable, as indicated by Charles Leland's poem, "The Pigeon," included in his 1876 book, *Pidgin-English: Sing-Song or Songs and Stories in the China-English Dialect.*[126]

Perhaps in addition to these theories the word "pidgin" could also be taken as a reference to the *carrier* pigeon's usefulness for secret communication, which hinges on the bird's ability to move messages among geographically remote parties. Indeed, if we look at another language's word for pidgin, we find an etymological circumstance similarly tempting such an association between pidgin communication and clandestine transportation. In English, the Chinese-English pidgins used on the docks of Shanghai in the nineteenth century wound up being called simply "Pidgin English." Ellen Semple distinguishes this Pidgin English, the "lingua franca in the ports of China and the Far East," from the West African and Caribbean pidginizations of English that are of interest to Linebaugh and Rediker.[127] In the Shanghai dialect of Chinese during the nineteenth century, the term for this pidgin argot used on the docks and quays of the city was the "Yangjing-bang" (洋泾浜) language.[128] This term, "Yangjingbang," was also the name of a canal that, just north of the old walled city, served as a divider between the British and French Concessions in New Shanghai from the 1840s on.[129] The canal has long since been filled in, but many people in Shanghai still use the term Yangjingbang to mean pidgin language (usually English-Chinese), or, more generally, a "poor imitation" of something.[130]

The Yangjingbang waterway itself seems to have been named for the long-standing presence of foreigners along the city's Huangpu docks. *Yang* (洋) was a term often used in the nineteenth century to connote foreignness or ocean trade. An 1860 English visitor to Shanghai translated "Yangjing-bang" as "Ocean Flowing Stream";[131] an 1889 English visitor called the canal the "Foreign Boundary Creek."[132] A historian of Shanghai notes that in the middle and late nineteenth century, it was along the Yangjingbang that "foreign and Chinese interaction was most intense, due to its proximity to the docks along the Huangpu River. The embankments of the shallow canal were popular sites for migrant shelters, shops, and brothels." Wang Tao, a late Qing reformist thinker, observed that "the Yangjingbang area is a floating world"—an overall impression perhaps also communicated by the presence of the Chinese water radical (氵) in all three characters forming the geographic name.[133]

During the Taiping Rebellion, the Yangjingbang played an important role as a space of subversive mobility and clandestine circulation among insurgent social groups in Shanghai. During the bloody uprising, Qing authorities wrote to American officials speaking of subversive waterfront elements' use of the Yangjingbang waterway to forward arms to Taiping rebels (the Small Swords Society) holed up in the old walled city during the fall of 1853. Wrote Jierhanga, governor of Jiangsu and vice president of the Qing board of war:

Whereas since Lin-Lee-chuen and other rebels of Fujian and Guangdong have clandestinely usurped the city of Shanghai, because Yangjingbang and its vicinity being a place where officers and merchants of all nations reside, we were afraid that shots of muskets and cannons might reach there.... Then there were traitorous parties of the inner land who privately concealed themselves in Yangjingbang, and viciously united with those wandering people of foreign nations to supply the rebels with provisions, gunpowder, cannons, and arms, so that though they have been besieged a long time, it has been impracticable to suppress them.[134]

Jierhanga goes on to praise the American minister of China, Robert McLane, for having agreed to cooperate with the Qing, British, and French authorities to build a new wall and mud embankment along the Yangjingbang, to cut the canal off from the insurgent supply lines. The praise, though, is slightly backhanded, since the very "wandering people" of the waterfront implicated in Jierhanga's letter were likely wayward American and British sailors, many of whose sympathies were with the Taiping rebels.[135] The impression was that the Yangjingbang—both the transport space and the pidgin talk—had enabled a clandestine intercourse to take place between linguistically dissimilar rebel factions: the insurgent "parties of the inner land" and the rebellious sailors from across the sea.

Unmappable Mobility and the Elements: Six Geographies of Possibility

I have considered six forms of transportation—by pigeon, mule, elephant, camel, sled dog, and watercraft—and the utility of these forms for the purposes of clandestine trafficking and insurgent mobility. One way of presenting the geographic parallel among these methods of transportation is to consider them as composing a kind of self-contained "color wheel" of transport through elements where fixed roads cannot easily go. The three "primary colors" on this wheel could be cast as pigeon mobility through the air, watercraft mobility in the water, and mule mobility on rocky terrain. By extension, the three "secondary colors" could be construed as the methods of transportation across more changeable, intermediary states of matter: sled dog transport across "rocklike water" (ice or permafrost), elephant mobility through "airlike water" (monsoon rains), and camel mobility through "airlike rock" (desert sands). H. Warington Smyth, an English engineer working for the Siam government during the 1890s, noted the parallel among these latter three modes of mobility: "What the camel is in the desert, and the dog upon the ice-floe, that is the elephant in the forests of Nan," a region in the northern Thai hills.[136]

Of course, such a framework is imperfect and rudimentary. It passes over many other vehicles, both animal and motorized, whose utility to rebel traffickers is certainly worthy of examination and study. But the framework is also analytically useful in that it helps to structure and define the broader topic, namely, the transport methods associated with politically subversive possibilities for mobility, not just according to the limited body of evidence presented in this work but also according to a wider and more comprehensive logic of transportative possibility across geophysical space.

We might imagine a country comprising many kinds of physical regions and ruled over by some occupying power. In this country the pigeon keepers, muleteers, dog mushers, camel drivers, water pilots, and mahouts have joined with each other into a kind of guerrilla confederation. Each form of insurgent transport labor has its niche in the overall network of covert conveyance. The pigeoneers can move lightweight cargo—secret messages, vials, computer chips, cell phones, and the like—nearly anywhere, provided that on one end of their avian charges' journeys the keepers establish secret coops that the airborne smugglers will know as home. The muleteers smuggle larger batches of rebel freight—perhaps mimeograph machines for turning out insurrectionary circulars—through remote mountain passes and through those parts of the temperate valleys that are still mostly forested (the occupying regime, endeavoring to bring as much of the country as possible under political and economic control, is hard at work conscripting labor to build regular roads through these forests wherever possible). The cameleers move arms, building supplies, and other awkwardly shaped manufactured goods across the dry, dune-swept margin. The dog mushers help political fugitives escape across the tundra and sea ice to hideaways beyond the occupiers' reach. The mahouts, of greatest use during the rainy season, guide rebel supply trains along rocky streambeds, through morasses of spongy mud, and across tidal shallows. Flotillas carry goods and people from the sea and up the navigable inland waterways. To finance the resistance, all six classes of transport labor also smuggle sellable contraband, such as drugs, alcohol, and exorbitantly tariffed items, through underground markets—the country's Wang Khas, Suakims and Yangjingbangs.

Within this clandestine rebel trafficking network, certain kinds of hand-off relationships and points of transshipment may become especially important. Che Guevara, in his treatise on guerrilla warfare, remarks on the desirability of partnering the rebel muleteers with keepers of carrier pigeons. In wet mountainous areas, the muleteers might also develop a relationship with the elephant drivers—much as how in Tucker's memoirs the Yunnanese muleteer, whose specialty is transmontane smuggling, has a hand-off relationship with the Kachin elephant drivers, whose specialty is shipment through the treacherous fords and steep, muddy *shat khats* of the Kachin Hills. The dog mushers may benefit from special knowledge of watercraft so as to maintain their routes during the season of thawing and ice breakup. Similarly, the mahouts and oozies, being of such importance during the season of floods, may develop a special relationship with boat-people. This likelihood may explain British officials' conviction, following the slaughter of their colonists at Cawnpore in 1857, that both the city mahouts and the Ganges River boatmen were to blame for the treachery leading to the garrison's destruction at the riverside ghat.

If the mountainous terrain favored by the muleteers is far from the open water, the network of rebel mule handlers may be unlikely to strike up a transshipment relationship with the waterborne smugglers. During the Greek Civil War, the muleteers of the Greek Democratic Army were mostly cut off from the sailors of the National People's Liberation Army, even though both of these forces were Communist-aligned. The potential for transshipment was negated by the remoteness of the Pindus range, favored by the mule drivers, from the haunts and bays of the Adriatic coast and Gulf of Corinth, which were favored by the rebel navy.[137]

Of course, no rebel confederation quite like this has ever existed, at least not within modern memory. Indeed, in the modern historical era, the only political force actually bringing all six of these forms of off-road mobility together seems to have been the British Empire during the late nineteenth and early twentieth centuries. During this period, the British Empire's mobile capabilities ranged from its seafaring navy—the connective tissue of the empire throughout its history—to its military camel cavalries, pack mule teams, and pigeon corps; from a dog-pulled police force in its North American Arctic domain to an array of elephant work crews on the plantations of Burma and Assam. It is curious, then, that the British were often hesitant to invest in and expand these transport capabilities, instead legislating against the Australian camel trains from the 1890s on; opting, in East Africa, for a human porter system rather than off-road capabilities organized around trained elephants; and divesting (as chapter 3 takes up) from the inland waterway networks of Britain and Canada. Such divestments from these unique capacities for off-road mobility reflected the increasing attraction of modern industrial capital to regularized motor transport. They also reflected a geographic retreat from the political possibilities associated with an older frontier.

Cartographic representations of space—maps or satellite photographs—are intrinsically useful for anticipating the movements of vehicles dependent on roads and tracks. Road maps communicate the fixed set of routes a wagon, car, or truck is able to follow, the possible paths that can be taken by a train, and the finite set of locations where a runway-dependent airplane can take off and land (seaplanes are another matter). Maps are less helpful for anticipating the movements of floating vessels or transport animals. Shelby Tucker's *Among Insurgents* ends with a "Cartographer's Note" by John C. Bartholomew, the president of the Royal Scottish Geographical Society during the early 1990s. On reading Tucker's 1989 diaries, Bartholomew agreed to help create a map of the likely route of Seng Hpung's guerrillas across the highlands of northern Burma. In his note, Bartholomew writes that "while modern air charts of the Kachin Hills"—he points to the 1:500,000 scale Tactical Pilotage Series of 1990 as an example—"had improved our picture of the north border, they had not enhanced the precision and detail of those parts of the Kachin Hills of concern to us." Bartholomew instead opts to trace Tucker's route using the British "Survey of India" topographic series, dating from the 1910s through 1940s. These older maps presented their own disadvantages. For instance, as a result of Kachin farmers' use of swidden agriculture, many clearings and villages of the early twentieth century had completely changed location by the 1980s. However, unlike with the aerial photographs, the early twentieth-century surveyors had painstakingly recorded information hidden by tree cover: the location of rocky fords, the direction of water flow, the presence of quagmires, portions of the gold-panners' paths, and so on.[138] And yet even these maps were not good enough to account for long stretches of the Kachin rebels' elephant-mounted march. As Bartholomew writes, "Those sectors travelled by petrol tankers and jeeps presented little difficulty, as the locations had not changed and were well documented.... The other sectors travelled by foot and elephants, however, presented a different order of challenge altogether."[139]

To the cartographer's eye, the route taken by an off-road pack team can be difficult or impossible to discern. With motorized off-road vehicles, the situation is often quite different. For instance, a single snowmobile moving across the tundra surface can, in effect, etch a longlasting "map" of its own journey into the frozen landscape. When a snowmobile moves across the tundra surface (making quite a racket as it goes), it transfers its energy downward and accelerates the thawing of the surficial permafrost.[140] This surface disruption in turn leaves a long chain of puddles in the vehicle's wake, as well as a significantly altered pattern of vascular plant and lichen growth relative to the surrounding ground cover. The linear chain of puddles and altered surface coloration can last for years, betraying the snowmobiler's route. Interested observers can find such snowmobile scars easily enough on publicly available satellite imagery. In Google's 2010 aerial set, such Arctic areas as Point Collie, Alaska, or the vicinity around Varandey, Russia, are criss-crossed with snowmobile tracks, likely dating (to judge by the tracks' varying shades) from many different past seasons. In the case of heavier off-road tundra vehicles, such as those used for seismic gas exploration, the linear impressions left by a single trip can linger for many decades. Indeed, large swaths of the Beaufort coast of Alaska, or of the Mackenzie River Delta of northern Canada, look from the air like a vast street grid of discolored vegetation.[141]

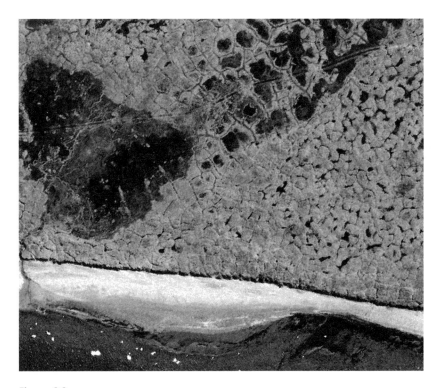

Figure 2.3
Aerial photograph of the coast of Point Collie, Alaska, from the summer of 2010 showing a snowmobile scar left from a previous winter's excursion. Such linear tracks in the permafrost landscape can last for several years, or, in the case of heavier gas exploration vehicles, several decades.
Source: Google Earth.

From the standpoint of a geographic analysis of subversive mobility, the point to take away from this cartographic legacy of many hundreds of thousands of snowmobile trips etched into the mappable landscape is that snowmobiles tend to give away their own route, and so are of rather limited use to smugglers—especially smugglers with an insurgent agenda. This limitation extends to most other off-road forms of motorized ground transport: ATVs, dirt bikes, jeeps, tanks, and so on. Except under very unusual geophysical circumstances—salt flats are sometimes solid enough to resist the formation of tire marks—these vehicles do not merely leave tracks, they substantially reshape the mappable features of the landscape in their wake.

By contrast, park rangers in Denali National Park, Alaska, have struggled with the opposite problem: of getting dog sleds to *leave* visible tracks in their wake, so that recreational mushers don't get lost in this northern wilderness. In other parks, the problem is easily solved by sending snowmobiles along the trails, but in Denali snowmobiles are banned because of the clamor produced by their engines. The snows around Denali are often of the light and dry variety that easily blows away—the nival equivalent, perhaps, of T. E. Lawrence's footstep-erasing windblown desert sands. Some of the rangers deal with the problem by attaching perforated sheets of metal to the backs of their dog sleds. Their hope is that, once "grated," the structurally altered snow will pack down into a durable and visible track as it might under the pressure of a snowmobile, rather than persist as a path-removing, dispersible state of matter.[142]

3   Fly-Boaters, Filibusters, and Canals

Previous chapters looked at the usefulness of animal transportation both for politically subversive logistical operations and for the filling of certain formal or "legitimate" economic niches in the modern era. One of these formal niches is the use of transport animals, such as camels and elephants, to aid in the geographic exploration and surveying of resource frontiers. Another is the use of transport animals to aid in relief efforts (such as the use of elephants during floods) when fixed infrastructure becomes unusable. And another niche still is the military utilization of pack animals for deployment in counter-guerrilla campaigns in countries with rough terrain and few roads. In many cases, such modern possibilities for animal transportation have gone unrealized.

The modern economic possibilities for water transport are far wider than they are for animal transport. Indeed, the infrastructure of water shipping, in the form of containerization ports, dredged channels for ships and barges, and global positioning networks for navigation, dominates global transportation in the twenty-first century. However, water transport infrastructure is also very geographically uneven. This unevenness is partly the result of preexisting physiographic conditions—some populated regions are simply too high and dry for shipping lane access to be a realistic proposition. But in many cases the unevenness of water transport infrastructure also stems from uneven human investments in areas where ship navigations could quite feasibly go.

For instance, far less modern waterway infrastructure serves the relatively flat, wet interiors of the British Isles or the Canadian plains than serves the interiors of Germany, France, the Low Countries, Russia, and the United States east of the Rocky Mountains. The relative sparseness of modern inland navigations in the British Isles and Canada is an outcome of British infrastructural capital having mostly opted out of the "ship canal–building era" that characterized most of the Western industrial economies between 1870 and 1920. During this period, British leaders—who in the late eighteenth and early nineteenth centuries had been world leaders in canal engineering—withheld financial support for almost all modern ship canal projects in the British Empire (the relatively short Manchester Ship Canal stands as an exception).

An important but often overlooked factor that militated against the realization of many of Britain's modern canal schemes was a pervasive political bias against canals and canal people, a bias that came to the fore once railroad technology appeared to provide an alternative means of shipment. To illustrate and provide a sense of context for such biases, I look at some of the associations, real and perceived, between canal people and exotic, "wandering" races; canal people and illicit activities such as smuggling, pilfering, and poaching; and canal people and subversive political activities such as piracy and anti-imperial agitation. These associations were in part facilitated by the physical spaces of the canals themselves, which enabled the clandestine movement of illicit materials and a social disjunction between the "floating" people of the inland waterways and the more stationary populations of the towns. Finally, to broaden this social analysis, I point to two anti-imperial political movements in which subversive agents succeeded in commandeering canal infrastructure in ways not possible with the railroads: the Ribbonist conspiracies between the 1840s and 1870s and the Fenian Raids of the 1860s.

Britain's Missing Ship Canal Era

The Black Country, which envelops Birmingham in the British West Midlands, has been called the "Venice of England," not for any basilicas or palazzos in its possession but for its many canals. These narrow waterways, known as the Birmingham Canal Navigations and dating back to the late eighteenth and early nineteenth centuries, wind their way through the undulating landscape, tunneling under ridges, moving across aqueducts, or climbing steep slopes by way of hand-powered lock-gates, of which there are sometimes a dozen in a row, leapfrogging each other up hillsides.

Most of the towns and villages of this inland countryside have, or once had, their own wharf facilities. The city of Birmingham still has its main canal lines and beautiful iron and wooden lock-gate structures, but the branch lines and city wharves, such as Old Wharf and the Baskerville Basin, the site of the infamous 1791 Joseph Priestley riots, are gone. James Vance, a late twentieth-century historical geographer, has written of his experience in 1960, walking from the eastern outskirts of Birmingham completely across that very large city to Smethwick in the west, always along the towpath of the Birmingham Canal Navigations. In that entire time I met not a soul on the towpath save one city garbage boat drawn by a single horse and steered by a lone elderly man at the tiller. For this period of five or six hours I hardly saw anyone as the canal crawled under streets, behind buildings with windows often bricked up, and always walled away from the currently used part of the city. In one place there was a slight suggestion of interest in the canal, as a door in a wall was open and a canvas hose was dropped into the green motionless water of the canal. The local fire brigade was using that nearly forgotten source to extinguish a fire. But no one was looking at the canal.[1]

At that time (since 1960 some English canals have been refurbished, if only for recreational use), this peculiar geography of negligence and abandonment extended beyond Birmingham and the Black Country to much of England. The British Midlands' Grand Union Canal, connecting London to Liverpool, was devoid of commercial or industrial traffic, as was London's famous Regent's Canal. In the Black Country between the 1850s and 1950s, more than half the canals were filled in, the water frontage lost. Similarly, in London, the Kensington, Grand Surrey, and Croydon Canals were abandoned and filled in.[2]

In many other advanced industrial countries the outcome has been very different. Though Britain may have abandoned its canals and canal-building projects after the mid-nineteenth century, the industrial powers of Europe, and to a lesser extent the United States and Russia, did not. In France and Germany in particular the late nineteenth and early twentieth centuries were a period of ambitious ship canal and barge canal building—a "shipping canal era," as one historian has put it[3]—during which political and financial leaders aggressively invested in wide, deep inland waterway facilities to complement those countries' burgeoning railway networks. Yet Britain, along with much of its empire, did not participate in this epoch of ship canal building. On twenty-first-century inland navigation maps, the contrast is striking: a European continental interior served by numerous deep-draft canal networks, all built during the late nineteenth and twentieth centuries, and a British interior served only by the narrow and shallow-draft networks realized during an earlier epoch.[4] The contrast begs the question of why. Why, after a long period of canal-building supremacy between roughly 1780 and 1869 (when the Suez Canal opened), did Britain wind up financing so few ship and barge canals compared with many rival industrial powers?

It wasn't for lack of ideas or geographic possibilities. Though the Suez Canal had originally been a French rather than an English project, British elites quickly recognized the economic value of the waterway for funneling Indian grain to British markets, thus driving down the costs of English factory labor.[5] Lancashire industrialists and Manchester elites were impressed enough with the hundred-mile-long, interoceanic canal at Suez that, starting in the 1870s, they began to push for a Manchester Ship Canal, a twenty-five-mile-long sea-level channel that would make Manchester a true seaport. The Manchester Ship Canal advocates managed to wrest the necessary legislation from Parliament and had the canal completed by 1894.[6]

This political and infrastructural success may well have launched a wider British ship canal era, and many contemporary observers, investors, and canal advocates certainly expected it to. Emboldened by the success of the Manchester seaport canal and its proven effectiveness in keeping down freight rates for the Lancashire region, many other British regional leaders between 1894 and 1920 began to show great interest in building wide, modern waterways for their own regions.[7] If only in newspapers, books, and reports (rather than in earth and stone), Britain experienced something of a canal renaissance during this time. Britain's waterway advocates began to point to other nations' waterway projects as models to learn from.[8]

This flurry of revived interest in canals culminated in 1909 with the publication of the *Report of the Royal Commission on Canals and Waterways*, which was chaired by members of Parliament from such inland cities as Birmingham, Coventry, and Nottingham, as well as by industrial leaders, engineers, and authorities on port management. The report discussed at great length the advantages enjoyed by Belgium, the Netherlands, France, and Germany, where modern waterways and railroads worked cooperatively with rather than competing against each other. Where cooperation existed, waterways handled heavy, bulky goods, such as grain and coal, while railroads handled smaller bulk goods for industries demanding higher rates of turnover.[9]

To canal advocates, the advantages of inland shipping waterways were clear. As one such advocate put it, "In the grand strategy of trade our competitors have acted upon the principle, well known to physics, that it is easier to move an object floating in liquid than to carry it on wheeled carriage."[10] The Royal Commission's report emphasized this view of canals as well, and suggested further that the building of wide ship canals across England could serve a social reform purpose, helping to spread industries and workers more evenly across the English landscape:

On a coastline, if full advantage is to be taken of the cheapness of sea transport, industries must be aggregated close to a few harbors. On a river or canal navigation they can be distributed along the whole line.... For social and sanitary reasons the distribution of industrial populations over wider areas is to be preferred to concentration in a few crowded districts.... If industries are widely distributed the workers can have better houses and lower rents, can breathe less vitiated air and they and their families can in many cases combine with factory work the healthy and profitable occupation of small agricultural production.[11]

Garden city advocates of the time, inspired by the ideas of the English social utopian Ebenezer Howard, recognized the potential significance of this reasoning for their own political movement. As the garden cities advocate J. S. Nettleford wrote in his 1914 *Garden Cities and Canals*:

Wharves and landing facilities can be erected at comparatively small expense anywhere along a canal bank, and therefore improved inland waterways are the best possible means of speeding up the garden city movement.[12]

Nettleford also viewed inland canals as special physical assets for the British working class. He pointed out that in other countries, such as Belgium, canal-building expenditure

has been largely undertaken at the instigation and demand of the working classes. This should encourage the British Labour Party to take an interest in the matter and not leave it in the hands of railways capitalists.[13]

The Royal Commission itself, though also partaking in such class rhetoric, was perhaps less concerned with the empowerment of British labor and more with the leaps and bounds in waterway building occurring in Germany, where the 1895 Kiel Canal had swung geo-strategic control of Baltic grain exports to German ports and where advanced industrial development was occurring along the Teltow Canal, the Finow Canal, and the Frankfurt-Mainz navigational network.[14]

To rival these German projects and assert Britain's preeminent position in international trade, the Royal Commission's report proposed a vast ship canal system for the whole of England. This system, referred to at times in the report as the "Cross," was to consist of two main shipping trunks, one linking Bristol to Kingston upon Hull and the other linking London to Liverpool; these trunks were to intersect at the inland cities of Birmingham and Nottingham. These new linkages were to bring Bristol and Liverpool closer to European markets and, it was expected, to facilitate industrial development in both London and the British Midlands.[15] Waterways were to handle a minimum standard of one hundred tons. The commission did further studies for a system in which the minimum standard was to be three hundred tons (the standard in many contemporary German and French projects was four hundred tons and up).[16] Construction and maintenance of the system were to be handled through private corporate conglomerates or through direct nationalization of the waterways of England and Wales. Political support for the Cross came, predictably, from the many regions and cities it was to serve: Gloucestershire, Leicestershire, Worcestershire, Lincolnshire, Warwickshire, Staffordshire, Northampton, Liverpool, Birmingham, Bristol, Manchester, Coventry, Cardiff, and Swansea. Numerous trade associations also supported the plan: dockers, builders, colliery owners, miners, and the Port of London Authority.[17]

Despite the massive amounts of ink and costly engineering studies that went into the formation and production of the plan (one historian even assesses the Royal Commission's report as "the most thorough enquiry made into any aspect of transport in Britain in the 20th century"[18]), the Cross scheme came to nothing. Several local factors can be highlighted to account for this outcome: Britain's status as an island, where a considerable number of the communities were coastal anyway and so would have had little interest in paying for inland waterways (though this would hardly explain the support of cities like Liverpool and Swansea for the scheme); the unfortunate timing of the plan, which came at a moment when Parliament was struggling with a political crisis in the House of Commons; and the onset of World War I.[19] Yet these explanations, focused on local predicaments and the "bad luck" of historical timing, would have considerable trouble accounting for the demise of comparable British ship canal schemes, not just in England but throughout the British Empire, and in particular in the British North Atlantic domain, from 1870 on.

For instance, a proposal in the 1880s by the eccentric British railway magnate Edward Watkin to build a ship canal straight across Ireland, from Dublin Bay to Galway Bay, passing through the inland city of Athlone, drew attention from magazines and newspapers at the time but was never taken up by financiers or political elites.[20] The proposal was engineered by the American James B. Eads, famous for his Mississippi Delta jetties (praised by Mark Twain in *Life on the Mississippi* as work "which seemed clearly impossible"[21]). Eads was also well known for his involvement in American efforts to build a ship canal across Nicaragua.[22] The Watkin scheme, if built, would have linked up with an older canal network, dating from the late eighteenth and early nineteenth centuries, that already wound its way through the eastern half of Ireland and was an important part of everyday life in the provinces of Leinster and Munster (whose river Blackwater some still refer to as "the Irish Rhine"[23]).

As with the 1909 Royal Commission's Cross scheme, Watkin's proposed Dublin-Galway ship canal came to nothing, and just as with the Cross, local, idiosyncratic explanations, such as British capitalists' hesitance to see

Ireland become industrialized, can be emphasized to account for this result. That such British bias against an Irish ship canal was soon after mirrored by British bias against *English* ship canals may be a simple coincidence. Yet if this is a coincidence, the coincidence extends even further westward, beyond the British Isles to Britain's vast North American domain, where multiple major ship canal schemes proposed between 1870 and 1910 also came to nothing.

One such Canadian proposal was for a ship canal across the Chignecto Isthmus, which separates the Bay of Fundy from the Gulf of St. Lawrence. The isthmus is only eighteen miles across and mostly low-lying and wet; plans to build a canal here date as far back as the seventeenth century, when the French-speaking Acadiens, who built small canals in their own towns along the shore of the Minas Basin, populated the region.[24] Interest in building a canal continued into the eighteenth and nineteenth centuries, by which time Britain had assumed control of the region and purged it of much of its Acadien population. A canal across the isthmus would have greatly shortened the distance from the St. Lawrence River to the Bay of Fundy—and so by extension from the Great Lakes to New England. Moreover, such a canal would have significantly shortened the trip between Liverpool and Saint John, New Brunswick—and Saint John, by virtue of its proximity to large reserves of timber, was in the mid-nineteenth century the largest shipbuilder in the British Empire outside Britain.

Numerous engineering proposals for a Chignecto Isthmus canal were penned during this period: in 1783, 1811, 1826 (in which the renowned English canal engineer Thomas Telford was involved), and 1843. Twenty formal surveys were made in all, up until 1875. In 1867, political leaders in New Brunswick entered into the Canadian Confederation with the understanding that elites in Quebec, Ontario, and Britain would help them finance a Chignecto ship canal.[25] Maritime leaders had a new Chignecto canal proposal ready two years later.[26] But these plans simply gathered dust—as did subsequent twentieth-century schemes, in 1928, 1931, and 1958. Indeed, in the twentieth century, the Canadian Maritimes had become so accustomed to this pattern, and so convinced of the Ontario and Quebec bureaucracy's inability to comprehend the transport needs of the Maritimes, that a New Brunswick song goes:

We've been part of this country a century or so
Seen promises come and seen promises go
But although Ottawa is the Maritimes' pal
They still haven't built the Chignecto Canal.[27]

Instead, infrastructural capital for the Maritimes was poured into rail linkages between Halifax and Quebec, and into the perhaps ill-conceived and never completed Chignecto Marine Railway. This unique engineering project, the brainchild of a New Brunswicker by the name of Robert Ketchum, was begun in 1888 and three quarters of the way completed before funding abruptly dried up in 1891.[28] The ship-rail consisted of two parallel locomotives designed to haul a massive cradle for carrying ocean-going vessels. At the rail terminus, ships were to enter a tidal basin and enclosed lifting dock and be positioned over a gridiron. The gridiron would then be raised by hydraulic rams and pressed to the track level. The locomotives would haul the ship at a speed of five to ten miles per hour, before arriving at the other end of the line and lowering the vessel back down into the sea.[29] Today it may be somewhat hard to believe that such a proposal was ever even considered, let alone three quarters of the way built, at massive expense (remnants are still visible today in the sparsely populated area around Sackville). Yet retrospective observers should remember what an important shipbuilding center Saint John was at that time, and the political fallout Ottawa risked if it could not point to some major Chignecto transportation project, as was promised to Maritimers in exchange for their commitment to the confederation.

The question remains, of course, as to why authorities did not instead use this thrust of capital for the often proposed ship canal. The unique physical geography of the Bay of Fundy, which at that time had the largest known tidal bore in the world, may have provided one rationale. Yet other, smaller projects in the Bay of Fundy region, such as the Shubenacadie Canal from Halifax to Minas Basin, had been successfully engineered before the 1870s.

Another historian has suggested that perhaps authorities in Ottawa did not want to realize for the Maritimes a real financial and commercial place in North American life; perhaps they were concerned about the monies that would pass through Saint John to the "Boston States."[30] This may be so, but, within the domain of the British Empire, a heavily studied ship canal proposal coming to nothing was not unique to Maritime Canada. As noted earlier, there were similar outcomes in the British Isles. Moreover, there were outcomes like this elsewhere in Canada, including within the watershed of the St. Lawrence, the traditional heart of Canadian power. In a sequence of expensive surveys, reports, and volumes reminiscent of the United Kingdom's *Report of the Royal Commission on Canals and Waterways*, a Canadian proposal emerged in the first decade of the 1900s to build a ship canal from Georgian Bay (an arm of Lake Huron) to Montreal by way of the Ottawa River. In comparison to the existing Lake Erie and Lake

SHIP IN TRANSIT

Figure 3.1
An idealized depiction, from around 1890, of Robert Ketchum's unfinished Chignecto Marine Railway.
Source: Courtesy of University of New Brunswick Archives.

Ontario route for getting goods from the North American Midwest to the North Atlantic, this inland waterway would have shortened by four hundred miles the shipment of goods from Georgian Bay, and by three hundred miles the shipment of goods from Lake Superior and Lake Michigan.[31] For commercial interests in the Midwest, central Ontario, Quebec, and Britain, the advantage of the route was obvious: it followed a relatively straight path from Lake Superior to Montreal, bringing the North American midwestern plains much closer to the North Atlantic and to Europe.

To American observers, the prospect of Canada actually building the Georgian Bay ship canal was so troubling that one American referred to the scheme as a "Banquo" to America's Macbeth:

The Banquo's ghost which will not down is Canada.... A ship canal would bring seagoing vessels to every lake port, through British territory, on a line so short that outgoing vessels taking it could be cleared from Montreal on their way to open sea before the Erie Canal barge starting at the same time and place could reach Cleveland.[32]

Another American observer voiced similar fears:

With a ship canal, however, which would permit the largest ocean steamers to go to and from the west end of Lake Superior, wheat could be grown in that vast region and be marketed in Europe at prices which would utterly destroy the wheat and corn markets of the United States.... When the British Northwest can raise and ship by canal 100,000,000 bushels of wheat, British capital will build the Georgian Bay Ship Canal, and every ton of traffic from the Lake Superior regions to the ocean will traverse British territory, leaving Detroit, Toledo, Cleveland, and Buffalo hundreds of miles from the direct route to the ocean.[33]

The engineering of a Georgian Bay ship canal, while expensive because of its length, was deemed by all observers to be well within the bounds of practicability. The longest stretch of the proposed waterway would have involved the canalization of the Ottawa River, whose course a *New York Times* article in 1895 had assessed as lending itself to becoming "one of the most perfect systems of inland navigation in the world. [The river] consists almost altogether of stretches of deep and still water, which are easily overcome by locks and dams." The major engineering challenge would be the creation of a cut from the upper Ottawa River to Lake Nippising (part of the Lake Huron watershed), about thirty-five miles away.[34] The challenge of building this cut was little, though, compared with the building of German canal cuts across the central German uplands, or with American engineering efforts in Central America occurring during this period.

The excitement in both Canada and the United States surrounding the prospect of a Georgian Bay ship canal was dissipated by the onset of World War I, after which point the canal plans were mostly forgotten, and the task of transporting Canadian grain and corn eastward was given to the Canada Pacific Railway, whose limited cargo space and difficult trek across the Canadian Shield made (and still make) for the relatively inefficient transport of bulk goods.[35]

The outcome was similar in the Canadian Midwest itself, where, between 1870 and 1910, dramatic waterway proposals emerged for the region: one for the canalization of the thousand-mile-long Saskatchewan River, which winds its way through the cereal wealth of the Canadian Plains and empties into Lake Winnipeg; another for a ship canal from Lake Winnipeg through

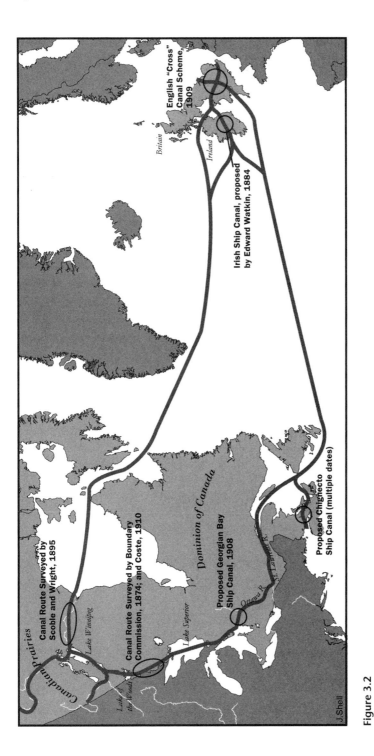

Figure 3.2

British North Atlantic ship canals had the potential to open up more direct shipping routes between the Canadian Midwest and markets in Great Britain.

Source: Azimuthal Equidistant projection centered on London. Cartography by J. Shell.

the Lake of the Woods to Lake Superior; and another for a seasonal canal from Lake Winnipeg northward, along the Hayes or Nelson River to Hudson Bay. Also proposed—though by American, not Canadian, authorities, during the American Civil War (and again in the 1890s)—was a steamship canal to link the Upper Mississippi River to the Red River of the North, which starts in Minnesota and empties into Lake Winnipeg.[36]

Much as the 1909 English Cross scheme proposed to do for the position of Birmingham within British transportation geography, such proposals for the Canadian Midwest had the potential to transform the Lake Winnipeg region into one of the most important shipping crossroads in North America, rather than simply a remote stop on the Canada Pacific Railway. As with the Georgian Bay ship canal, American observers thought the advantages of the Winnipeg canals to be so obvious that the United States would have to alter its own transportation policies to adjust to their northern neighbor's coming transport revolution.

For instance, in 1895 the *New York Times* suggested that Canada's coming "Waterway of the North," linking the Red River to Hudson Bay, was going to pry the shipping interests of the American upper midwestern states away from the interests of the Great Lakes and East Coast states and encourage them instead to look northward, to Manitoba and Saskatchewan, for common shipping aims. The article then pointed to the plans of Winnipeg's Lieutenant Colonel Thomas Scoble and Archibald Wright, who had surveyed several routes earlier that year to link Winnipeg to Hudson Bay by canal, with a seaport at York Factory or Port Nelson. Anticipating the possible southern bias against building major works of infrastructure in the Far North, the *New York Times* asserted that "the climatic bugbear should not deter the people of the Northwest from making use of their waterways. Russia, under much less favorable conditions, is doing wonders with her arctic seaports," such as Archangel, on the White Sea. As the article notes, due to the curvature of Earth, the Scoble and Wright proposal would have brought Winnipeg closer by ship to Liverpool than is Chicago—despite Winnipeg's location hundreds of miles west of Chicago.[37]

Lawrence Burpee, a Canadian civil servant and writer for *Popular Science*, further emphasized the possible significance of such a canal to the north, noting that the Hudson Bay route would save Winnipeg six hundred miles, Duluth five hundred miles, and Saint Paul one hundred fifty miles in the shipment of produce to Europe. Burpee also pointed to the findings of a recent Arctic expedition, that of the *Neptune* in 1903 and 1904, as evidence that the Hudson Straits were "safely navigable for long enough each year to make the Hudson Bay route commercially successful."[38]

The American writer Herbert Quick adopted a similarly excited tone with reference to the prospect of Canada building Winnipeg canals. His evocative vision of the northern transportation geography to come is worth quoting at some length:

What moves are open to Winnipeg? Two important ones: she can go to the sea by way of the Nelson River from Lake Winnipeg to Hudson's Bay, or she can come into Lake Superior with any sort of canal she desires, or—and this is what she will do— she can do both.... The great feeder of Lake Winnipeg from the east is the Winnipeg River, rising in the Lake of the Woods, on the northern border of Minnesota.... From Lake of the Woods extends an uninterrupted series of waterways ... a magnificent series of connected stretches of open water fit to carry the commerce of Winnipeg's mighty young empire down to Thunder Bay, on Lake Superior, with only a few miles of canal. There are some rapids, some falls; but the engineering difficulties are as nothing compared with the objects to be accomplished. The very logic of the earth's surface, the demands of an abounding commercial ambition and the power of abounding wealth must finally demand the making of this great way: and we shall see these beautiful lakes and rivers among the great commercial waterways of the world. Here must be the Erie Canal of the new West, or the ship canal to its Chicago. The waterway system of Canada will then have become coextensive with her productive area, and ships may load from Winnipeg's docks for all parts of the world.[39]

Nor, as noted by another commentator of the period, a geographer by the name of Watson Griffin, did such a vision need to end at the region around Winnipeg. In an 1890 article for the American Geographical Society of New York, Griffin went so far as to postulate that this projected waterway system of the "new West" could be extended even farther up the western interior lowlands of the continent, into the northward-flowing basin of the Mackenzie River. Griffin commented on what he saw as the potential interhemispheric significance of such a linkage:

When Champlain first reached the upper end of Montreal Island and saw the Ottawa River he exclaimed: "*La Chine!* This is the way to China!" That is why we call our Montreal Canal "Lachine." The name will be justified when the St. Lawrence, Nelson, and Mackenzie systems of navigation are connected, for then a small steamer leaving Montreal harbor will pass through the Lachine Canal, up the Ottawa to Georgian Bay and by way of Port Arthur, Lake Winnipeg and the Saskatchewan to the Mackenzie, which it will navigate to the Arctic Ocean, where a larger steamer will be waiting to take the passengers through Bering Strait and across the Pacific to China.[40]

For Griffin, the immediate utility of this western interior canal corridor would be to support imports from the Beaufort Sea whaling industry, at its economic height during this time. Griffin also envisioned that the outward voyage, via the Mackenzie Delta to East Asia—approximating the Great Circle that runs from the North American interior grasslands to the Yellow Sea—might be "made occasionally by adventurers"; he did not comment on which bulk resources abundant in North America's interior (coal, wheat, and so forth) would be, in all likelihood, much prized by industrialists in Japan.[41]

It is perhaps rather striking that one encounters these sorts of bold and evocative assessments of Canada's potential for future waterway development in many American but few Canadian publications of this period— even though, to be sure, the actual surveys of possible canal routes were all from Canadian officers and engineers, such as Scoble and Wright along the Nelson and Hayes Rivers, the surveys of a British International Boundary Commission during the 1870s for a possible all-Canadian canal route from Lake Superior to Lake of the Woods,[42] and the efforts of Louis Coste, a former deputy minister of public works, to survey a Lake of the Woods canal in 1910. This final surveyor, Coste, contended that "sooner or later such a canal would be demanded by the country, not only along this route but from Lake Superior to Edmonton. Otherwise ... Canada will lose the grain carrying trade of the country."[43] Yet if in the United States, ship canal proposals were frequently headline material for the *New York Times*, in Canada, Coste's report was overlooked by the *Montreal Daily Star*.

One explanation for this gap in attitude may be the relative strength, during that time, of the American publishing industry. American readerships, including those along the major shipping channels, eagerly devoured publications like *Harper's*, *Van Nostrand's Engineering Magazine*, and *Popular Science*, whereas Canadian readers tended to focus more on the writing output from Britain. Another explanation may be that some Americans were somewhat confused about the actual waterways and geography of Canada. For instance, one misinformed Iowan at the 1909 Trans-Mississippi Conference stated that Canadians,

with less than one twelfth of our population and with wealth absolutely insignificant to ours ... have built the Georgian Bay Canal.... They are also constructing a ship channel from Thunder Bay ... to Winnipeg in Manitoba. This is deep water nearly all the way.... They are also constructing a ship channel from Winnipeg through Lake Winnipeg, which is as large as Lake Erie, Nelson River to Hudson Bay.... They have also constructed a canal from Winnipeg into the great wheat country more than 1000 miles and in the other direction more than 500 miles.[44]

Needless to say, Canadians hadn't done any of these things. But Americans' inaccurate image of Canada as an energetic waterway builder may have done much to provoke their impassioned speech in regard to waterways and to fan the political flames that successfully agitated between the 1890s and 1920s for major ship canal projects in Illinois, Massachusetts, New York State, Maryland, Texas, and Louisiana.

Nonetheless, even if Canadian publications of that period were often rather dispassionate or dismissive in their assessment of the major Canadian ship canal proposals, later Canadian writing suggests that, among the transplants from the East who settled the Canadian Midwest in the late nineteenth and early twentieth centuries, there was always a tremor of excitement at the prospect of setting sail from places like Winnipeg, Saskatoon, and Moose Jaw to the ports of the world. The popular Canadian writer Farley Mowat, who grew up in Saskatoon during the years after the Great Depression, wrote in his 1957 novel *The Dog Who Wouldn't Be* of the comically futile attempts of his father, Angus, to get boats down the flood-prone Saskatchewan River to Lake Winnipeg and beyond.

In the novel, Angus Mowat is a native of the Bay of Quinte, an arm of Lake Ontario linking the lake to the Trent canal system of southern Ontario. He has left the province, having grown "tired of the physical and mental confines of a province grown staid and stolid in its years."[45] He "came by his passion for the water honestly."[46] In one chapter, Angus and another Saskatoon local by the name of Aaron Poole concoct a scheme to build a seaworthy vessel, dubbed *The Coot*, right there in Aaron's basement, and sail it to the Gulf of Mexico and then up the Atlantic coast to the Gulf of Saint Lawrence. Mowat, speaking perhaps for a later generation that had already adjusted to landlocked life on the prairies and considered this "longing for the sea" to be an old man's folly, calls this route "one of the most unusual ever attempted, not excluding Captain Cook's circumnavigation of the globe."[47]

As for his father's friend Aaron, Mowat makes some fun of him as well: he had "originally come from the interior of New Brunswick, and had never actually been to the sea in anything larger than a rowboat during his maritime years." But this, Mowat says, was "not relevant to the way [Aaron] felt.... As a Maritimer, exiled on the prairies, he believed himself to be of one blood with the famous seamen of the North Atlantic ports."[48]

The prospect of Angus and Aaron's voyage excites many Saskatoonians, though not the local railway officials. Mowat describes how the Saskatoon Chamber of Commerce hailed the venture with the optimism common to such organizations, predicting that this was the "Trail-blazer step that would lead to Great Fleets of Cargo Barges using Mother Saskatchewan to carry Her Children's Grain to the Markets of the World." On the other hand, the officials of the two railroads made mockery of *The Coot*, refusing to accept her as a competitive threat in the lucrative grain-carrying business.[49]

The Coot's maiden voyage is a predictable fiasco. After a crowd of Saskatoon townspeople see *The Coot* set off from the banks of the Saskatchewan River with Angus, Aaron, a horse, and a dog on board, they expect to hear reports of the vessel's progress from the next town downstream, some thirty miles away. They hear nothing, and an air of suspense sets in. Then the townspeople begin to hear peculiar reports from a village fifteen miles downstream, of a Ukrainian woman who saw a hull being hauled "by—she crossed herself—a horse and a dog. It was accompanied, she continued, by two nude and prancing figures that might conceivably have been human, but were more likely devils. Water devils, she added after a moment's thought."[50] The hull of *The Coot* had sprung so many leaks that Angus and Aaron had given up bailing water from the hull and instead carried the vessel along the riverbank, work so muddy and exhausting they'd resorted to removing their clothes, appearances be damned.

Such was these Canadian Midwesterners' thirst for the amenity of water navigation.

Railroads versus Canals

To be sure, histories of these various inland waterway and ship canal schemes in Britain, Ireland, and Canada between 1870 and 1920 can be written with *local* factors emphasized for understanding how so much surveying was done, so much engineering expertise consulted, so much ink expended on reports, yet so little actual earth dug. But the pattern is not local; it is exceedingly likely that more widespread and systemic forces were also at play.

In this section I look at several factors stressed, to varying degrees, by some other observers of and commentators on the subject, for understanding the early demise of canal investments in the British-controlled domain. These factors include the supposedly dissatisfactory technical performance of canals compared with railroads, the supposed superiority of railroads compared with canals for the purpose of creating national markets, and what might loosely be called "psychological" attitudes toward railroads as opposed to canals as objects of cultural fascination.

I examine the first possibility, that the decline of canals in the British domain is explained by the transport mode's supposedly weak technical performance, less as a response to any specific scholarly argument—in fact, most scholars have deemphasized the historical importance of a technical performance gap between canals and railroads[51]—and more to straightaway address what is perhaps the most commonly assumed explanation for the demise of nineteenth-century canal building. This "ready-made" explanation once dispensed with, the discussion then turns to more intriguing explanatory approaches.

Certainly, the technical limitations of canals compared with railroads were to some extent a factor militating against British canal investment, at least in certain areas. Conveyance by canal was, after all, quite slow, rarely exceeding a few miles per hour on narrow canals or ten miles per hour on wide, modern ship canals. To industries dependent on factors of speed, rail transport represented a clear improvement over shipping by canal. Moreover, in some regions, canals were frozen for many months of the year and needed to be either de-iced or left seasonally dormant. Finally, compared with railroads, canals were relatively constrained by inland physical geography.

Yet most historians who have closely examined the economic histories of canals and railroads during the late nineteenth century have deemphasized the importance of these technical differences. For instance, in his 1964 *Railroads and American Economic Growth: Essays in Econometric History*, the economic historian Robert Fogel, taking issue with W. W. Rostow's earlier thesis that the railroad had "triggered" an American economic "take-off" during the late nineteenth century, launches the intriguing counterfactual argument, namely, that even had no American railways ever been built, an expanded American canal and waterway network would have been able to handle most if not all of the late nineteenth-century American trade that instead came to be handled by the railways. Fogel makes this argument through a quantitative analysis of the late nineteenth-century American river and canal network, expanded to include thirty-seven canal navigations that were proposed but never built. Scrutinizing this "system of canals that could have been built in the absence of railroads," Fogel found that all but 7 percent of the agricultural land in the eastern United States would have been within forty miles of a water landing.[52] Fogel also incorporates the problems of speed and winter freezing into his analysis. He points out that for the staple commodities driving North American economic expansion during the nineteenth century—grain, coal, metal ores, sugar, and cotton—speed would have been a relatively marginal advantage, whereas rail transport's inability to replicate canals' bulk capacity was a marked *dis*advantage for the railroads. Fogel goes on to point out that freezing was not a problem in southern waters. For the northern canals, Fogel's scenario assumes the likelihood—given the absence of a rail alternative—of earlier and more widespread improvements in ice-melting technologies for keeping canals open during the winter months. Fogel points to a relevant 1873 proposal from the engineer R. H. Thurston to pipe artificial heat along the Erie Canal.[53]

Fogel thus concludes that the "social savings" from a counterfactual late nineteenth-century American transport system devoid of railroads and structured entirely around waterway navigation would have approximated or equaled the actual social savings afforded by the railroad-dominated transport system that actually emerged. In other words, the railroad was "a part of rather than a condition for the industrial revolution," and a full account of the financial "mania" for railroads during the late nineteenth century—not just in the United States but also in Britain, Canada, and many other places—must begin by looking at political and cultural context rather than at canals' technical flaws.[54]

Other economic historians, though often using different econometric methods, have extended Fogel's analysis to nineteenth-century Britain[55] and to nineteenth-century Canada[56] and have similarly concluded that the social savings from railroads in these places in the late nineteenth century were not terribly different from the savings derived from water transport. Unfortunately, such historians have yet to extend the counterfactual dimension of Fogel's argument to these places—that is, to look not only at real savings from existing canals but also at the potential savings from canals that were proposed but never realized.

There are, to be sure, some problems in the Fogel analysis. For one thing, Fogel's quantitative comparison of railroad mileage with canal mileage is problematic, since, in elaborate waterway networks involving not just rivers and canals but also lakes and bays, the total "mileage" of all potential water routes, including ferry crossings, linkages among lake piers, potential zigzagging routes, and so on, would surely approach infinity. In other words, the quantitative comparison with railroads, a transportation technology whose layout and usage are more circumscribed, is misleading.[57] Also problematic is Fogel's notion of social savings, a relatively stable and useful concept for weighing technical performance factors but much knottier in socially and politically mediated realms, such as the establishment of labor prices.[58]

Despite these analytical difficulties, Fogel's technical performance comparison of railroads and canals in the nineteenth century, stressing the extent to which canals' technical disadvantages were muted or compensated for by their innate advantages, is compelling, and certainly helps to explain why so many countries—including the United States, a point Fogel does not emphasize enough—outside the British Empire went to considerable lengths in the late nineteenth and early twentieth centuries to develop extensive inland ship and barge canal networks.

Railroads and the Creation of National Markets

A more sophisticated approach for explaining weakness in canal investment as compared with railroad investment in the British domain in the late nineteenth and early twentieth centuries stresses that insofar as railroads were more spatially flexible (at least over land), they were better at promoting national markets. Many analysts, recognizing the compelling technical advantages of canals and searching for a political economic explanation for their demise, have adopted this "national markets"–oriented perspective. Roy Wolfe, for instance, assesses the railroad as having been the great "nation builder" that drew together the industrial regions of Britain and brought Halifax closer to Ontario.[59] Charles Hadfield similarly points to the "narrow limits of the commercial as well as physical framework" of inland waterway systems, which are significantly determined by the physical geography of rivers and aquifers.[60] James Vance stresses this explanation as well, speaking of Britain's late nineteenth-century need to "integrate" its economy as a major reason why the geographically constrained canals lost investment.[61]

Philip Bagwell and Peter Lyth are even more emphatic on this point, contending that "it is hard to see the British canal age as representing anything approaching a technological breakthrough or the basis of a national transport system; certainly it is not comparable to the Railway Age of the mid-nineteenth century.... The idea that canal excavation was little more than an inevitable step beyond river improvement is persuasive."[62] These authors posit that canals were well suited for regional transport tasks like the shipment of coal to nearby industries, but not well suited for tasks like moving agricultural produce drawn from a wide area. Similarly, the distribution of manufactured goods did not lend itself so easily to canal transport, as final destination points were often widely spread over an entire national territory.[63]

The early twentieth-century canal advocate Herbert Quick also recognized this apparent advantage enjoyed by the railroads. By virtue of "being everywhere," railroads were better able to shore up contracts and political allies at a national scale. Quick further elaborates on this idea, in his characteristically evocative language:

The waterway is passive, lying in readiness to receive freight, but making no effort to get it, sending out no solicitors, pulling no traffic wires; but the railway is active, pervading the business life of the community, looking out for itself, doing favors, cutting to the bone in the waterway's narrower field, and making its losses up on something else.... The waterway is a logy animal of almost incalculable strength, but of low organism, and unprovided with brains; the railway is highly organized, efficient in brain, and knows just how to bring its strength to bear on its antagonist's weaknesses and limitations.[64]

This strand of political economy argument, stressing the political advantages enjoyed by railroads for the purposes of creating national markets, is compelling. Certainly, it cannot be dismissed in the search for factors that may have precluded the emergence of any "canal moguls" to match the emergence of railway moguls in the British realm during this period. Whether this explanation deserves analytical primacy, though, is another question. It suffers from three major weaknesses. One, it does not address why the canal-building era continued unabated outside the British domain. Two, it does not adequately deal with the possibility that some "nations" (such as, some might say, the United Kingdom) are geographic archipelagoes, the political and economic integration of which demands boats rather than trains. Three, and much more serious, this political economy argument doesn't match up with the economic history of the period, least of all for the British domain. The period from 1870 to 1914 wasn't a "nationalist" period in the development of capitalism but rather a period oriented toward deregulated international markets, with cross-national ownership of securities reaching extremely high levels compared with what was to come during the mid-twentieth century.[65]

For Britain, the crucial shift toward this integration of international markets occurred in the late 1840s, with the repeal of the Corn Laws in 1846 and of the Navigation Laws in 1849. Up until this point, both of these regulations had privileged the import of goods from British-held territories. The annulment of the Navigation Laws sparked a riot in Montreal, where Parliament buildings were burned, and merchants began agitating to join the United States.[66] Yet however much it might have weakened British colonials' feelings of loyalty to the British crown, the repeal of the Corn and Navigation Laws was also a central part of British capitalists' new import strategy, on the heels of the Irish potato famine, to reduce the cost of food through cheap foreign imports, thus—it was hoped—reducing the cost of maintaining labor power in the British industrial heartland.[67] Especially attractive to British advocates of this new "laissez-faire" import system were cereals from Russia, where costs were driven down by peasant labor, and cereals from the American plains, where the modernization of agricultural machinery had driven down grain prices.

Yet, as economic historian Deepak Nayyar has pointed out, this laissez-faire period was more than just an adjustment in trade laws, it also reflected a revolution in transportation. Nayyar observes that, during the second half of the nineteenth century, the substitution of steam for sails and iron for wooden hulls reduced the cost of ocean freight by two-thirds.[68] Similarly, the opening of the Suez Canal in 1869 halved the distance freight had to travel to get from Bombay to London. Britain's late nineteenth-century hunt for cheaper grain sources was thus vitally dependent on the existence of a new global system of shipping channels. In this sense the laissez-faire economic period is more revealingly understood as the "ship canal" *transportation period*—for these ship canals, on which the patterns of apparently "free" international trade flows depended, were built not through any laissez-faire decision-making process but rather arose within a web of geopolitical rivalries and stratagem. Barge canal projects in Russia and Ukraine to smooth the flow of Russian goods to the Black Sea were financed in large part by French capitalists to undercut British access to Russian grain.[69] Tsar Nicholas II, wishing to play the two Western powers against each other, financed the Mariinski and Kronstadt waterway modernizations to help ease the flow of Russian grain to the Baltic and to Britain.[70] Germany financed its own Kiel Canal to assert control over the Baltic grain trade and to create a ship link for its warships to move between Baltic and North Sea German ports.[71]

Table 3.1
British capital exports for waterways and railroad projects, 1865–1914

Capital Export Destination	A: Capital called for Railroads (in £000)	B: Capital called for Waterways (in £000)	Ratio A/B
Egypt	1,378	4,240	0.3
New Zealand	1,712	3,242	0.5
Australia	3,504	1,603	2.1
Germany	1,012	203	5.0
Cuba	12,863	2,501	5.1
Austria-Hungary	2,449	425	5.8
Turkey (Ottoman)	9,465	1,191	7.9
Brazil	54,553	1,158	47.1
France	1,219	23	53.0
Poland, Low Countries, Scandinavia	10,450	187	55.9
Global export of capital	*1,294,005*	*20,807*	*62.1*
India	128,376	1,275	100.7
Argentina	200,937	869	231.2
United States	514,837	890	578.4
Canada	166,337	277	600.5

Source: Raw data from Irving Stone, *The Global Export of Capital from Great Britain, 1865–1914: A Statistical Survey* (Basingstoke, UK: Palgrave Macmillan, 1999).

Especially revealing during this period are the choices of British financiers regarding the export of British capital to canal and dock, as opposed
to railroad, infrastructure projects abroad. Other than in Egypt (where the
Suez Canal dominated infrastructure building) and New Zealand and Australia (where new ports had to be constructed from scratch), British capitalists financed railroads much more heavily than they financed canal and
dock projects.[72] This is not terribly surprising in and of itself; rather, what
is revealing is a comparison of the ratio, for each capital export destination
country, of railroad capital exports to waterway capital exports. For such
capital export destinations as Germany, Austria-Hungary, Turkey, and Brazil, this ratio was relatively low, meaning British capitalists were funneling
comparable amounts of capital into both railroad and waterway projects. In
France and some North Sea and Baltic nations, the ratio was higher, meaning British capitalists' interest in railroads in those areas was relatively high
compared with their interest in waterways, but the ratio was still below the
international average.

Above the international average was Britain's largest Asian territory,
India, as was Argentina. Well beyond this was the United States; and higher
still was Canada, into which British capitalists were pouring *six hundred
times* more railroad capital than they were waterway capital. This figure is
all the more extraordinary when we consider that, unlike countries such as
Russia (where canal projects were financed predominantly by French capitalists and the tsarist regime rather than by British financiers), Germany,
France, and the United States, Canada had virtually nowhere else to turn
other than the London banks for large sums of construction capital. It is
also surprising in light of the substantial cereal wealth of the Canadian
prairies, which if cheapened through appropriate transportation investments might have competed with and driven down the prices of cereals
from India, South America, Russia, and the Baltic. The effect of this reduction in grain prices would have been a reduction in the cost of British labor
power. Yet British capitalists chose not to forward to Canada the waterway
financing they so happily forked over to many other nations, including
many political rivals.

What is also striking during this supposedly free-market, laissez-faire
period is that investment in easterly and southerly shipping routes coming into Britain from the Baltic, Mediterranean, Indian Ocean, and South
Atlantic, as opposed to westerly shipping routes coming in from the North
Atlantic—from Ireland, Canada, and the United States—coincided with
unprecedented central planning efforts to turn London, whose port faces
east and south, into a viable modern seaport capable of rivaling or eclipsing

Liverpool, which faces west. In Liverpool, the major additions to dock facilities during this period included additions to the Albert Dock. However, by the turn of the century the Albert Dock, designed to handle sailing vessels, had fallen into disuse.[73] The construction of the Canada Dock in 1859 was also a significant addition to the Port of Liverpool for handling incoming timber.

Changes in the Docklands of the East End of London were much more dramatic. In the early nineteenth century the Port of London was of less importance than the Port of Liverpool. London was England's preeminent commercial and political center, but not its center for freight handling or industrial activity. Yet the construction of the London Royal Docks between 1855 and 1921 provided London with facilities dwarfing those in Liverpool. The Victoria and Albert Docks, built by the London & St. Katharine Docks Company, provided large berths for seagoing vessels that could not be accommodated farther upriver. A great commercial success, these docks specialized in the unloading of cereals and other foodstuffs, with rows of giant granaries being sited alongside the quays. Complementing the expansion of the Port of London were dramatic alterations in the fabric of the city, such as the construction of Tower Bridge.[74]

The late nineteenth and early twentieth centuries may have been in some sense an important period for economic nation building, in particular in the United States, which was developing an industrial heartland of its own and had a vested stake in creating viable transport routes across the Rocky Mountains. Yet this period was also an important one for getting goods out on the open seas, and ship canals, where practicable, were a more direct way of doing this than railroads. From the standpoint of overseas trade, the period between 1870 and 1920 was the ship canal era first, and the railroad era second.

Yet the economic data suggest that British capitalists had a peculiar geographic bias during this period about where to invest in ship canals—a bias that had as little to do with the aims of asserting a "national market" as it had to do with any so-called laissez-faire policy toward imports. This bias, against points to the west and favoring the east and south, was geopolitical. The nature of this particular geopolitics bears closer examination.

Railroads as Psychological Symbol

Some accounts from the time point to another possible explanation for the demise of canals and canal building in the British domain—not technical or political economic but rather aesthetic and psychological. As a pamphleteer of the late 1840s noted, there was "an air of assumption and parade about a railway" that dazzled and deceived the superficial observer. It was "the very type of enterprise that embodied energy and efficiency," whereas the canal appeared to be "the embodiment of quiet, plodding, undisguised sluggishness."[75] Transport historian W. T. Jackman elaborated on this perceptual difference in 1916:

The canal traffic was carried on comparatively quietly. The sight of an occasional horse passing through the country, mounted or driven by a boy, and hauling an insignificant looking barge which was managed by one or two persons was no surprise to anyone.… On the other hand, the railway had an appearance of grandeur and ostentation that charmed the public.[76]

Jackman also points to a marked difference in attitude and culture between the canal proprietors and the railway men. The former was more an "old man's" line of work, one whose practitioners by 1850 were already characterized by pessimism and gloom. Railroad owners, by contrast, were "younger men, usually of the trading and industrial classes, who were actively pushing the construction of railways." Jackman further suggests that the canal owners had given up the battle with the railways "in despair, and perhaps at too early a period, before they had learned what strength they really had and how largely the traffic of the country would increase."[77]

These sorts of explanations, focused on people's attitudes and psychological biases, may get closer to the heart of what happened to the waterways of the British domain during the late nineteenth and early twentieth centuries. For these attitudes reflected not just a psychological bias against the canals themselves but also—and, I contend, of more importance—against the *people* who worked and drifted along these canals' courses. Technical and political economic explanations for the demise of inland waterways in the British domain are in many cases valuable, providing an important (if by itself insufficient) dimension to the picture of the British Empire's transportation history. Yet far too much is missed if we do not attend to the social history of the canals, centered now on the canals' "boatpeople," the political troubles they sometimes facilitated, and the peculiar fears they inspired. It is to this history, of the specter of canal-borne subversive mobility, and of political biases directed against canalling people, that I now turn.

Canal People

An English writer by the name of Ernest James Oldmeadow published the novel *Coggin* in 1920. The story is set in a fictional English canal town called Bulford-on-Deme, and opens in the year 1848. The scene is a canalside public house called The Barge Aground, which is, in the words of a local rector, "a hotbed of atheism, Chartism, and every other form of infidelity and treason," attracting "every type of Liberal, not excepting Socialists and Republicans."[78] This pub is abuzz with news of the revolutions in Europe. Patrons excitedly exchange rumors and tidbits of information regarding the possible assassination of the Duke of Wellington, the progress and travails of the Chartist leader Feargus O'Conner, and the looming collapse of many monarchies on the European continent. George Placker, the principal character of the book, is introduced as one of the patrons of The Barge Aground—a quayside painter whom Oldmeadow describes as "not by nature a rebel, or even a malcontent, [but] his blood had warmed and his nerves had tingled in sympathy with the spirit of revolution which had so lately overturned Louis Philippe and shaken many a throne."[79] The subsequent storyline involves Placker's efforts in the years following England's failed '48 revolts to educate a Bulford-on-Deme working-class boy, Harry Coggin, the talented son of a local canalside junk dealer. Placker's determination in this goal stems from the prospect of the young boy winning a local scholarship, historically awarded only to the children of Bulford's bourgeoisie. These main events of the story occur in the early 1850s, and by this time Placker has become more radical than he was in 1848. He has taught himself to read and write properly. And he has found his political voice. When his young protégé Coggin wins the scholarship, Placker urges the patrons of The Barge Aground to march through a wealthy section of the town, Victoria Park, "the new and expensive suburb, inhabited by Bulford's most successful burgesses, who were hated and despised by Placker's faction."[80] Oldmeadow describes this fictional working-class march on a bourgeois English suburb with some relish:

The irruption of the Plackerites into Victoria Park provoked a small panic.... When pandemonium, blaring and blazing, was let loose in Albert Crescent, the inhabitants thought that the end of all things had come. For two or three years the more doleful park-dwellers had been foretelling a revolution. Louis Philippe had lost his throne, and more than one dynasty had fallen. Why and how should England escape?... News of the coming invasion had reached Bridge Street nearly half an hour before, and the wildest rumors were flying about. It was said that five hundred desperate men from Demehaven, mostly Chartists, had been smuggled up the canal in barges and that they had sworn to seize and burn Bulford Town Hall.[81]

The canal-borne Chartist forces never come, of course. Placker's intent had been to use the scholarship to flame the political energies and self-esteem of the local working class; Oldmeadow's, in penning this tale, is to reveal the deep-seated fears and feelings of fragility of the bourgeoisie in a small canalside town in mid-nineteenth-century England. Such fears centered on the revolutionary events that had swept Europe just a few years before; and on radical political parties like the Chartists and Socialists, with their charismatic leaders and their propensity to riot. But, as the narrative construction of *Coggin* indicates, such fears also centered on the canals themselves, and on the people who worked on the canals and along their banks.

As the English canal historian Charles Hadfield has noted, Britain's early and mid-nineteenth-century canal population was marginalized and abhorred during that time: "no Engels studied them as he studied the condition of the working classes in Manchester in 1844; no Royal Commission investigated hours or conditions; no Charles Kingsley took canal people for his theme." Now and then, as this same historian notes, a glimpse of the canal people's lives appears in the records, filtered through a decidedly negative lens. Glimpses of the part supposedly played by canal people in spreading cholera, as for instance at Braunston, where the disease first attacked a woman washing the bedding of a canal boatman. Glimpses of the canal people's role in poaching, as when a Cromford Canal by-law prohibited any boat from stopping at night in Crich Chase, or in any wood or coppice. Glimpses of their role in encouraging drunkenness, as for instance requirements on the Staffordshire and Worcestershire Canal that lockkeepers report drunken pilots to the police authorities.[82]

Farley Mowat, whose Ukrainian Saskatchewanian woman of *The Dog Who Wouldn't Be* calls two hapless Saskatchewan River boatmen "water devils," may have been appealing to (while also parodying) a more longstanding feeling of moralistic hostility toward inland boatpeople. An 1818 reference to England's canal and river boatmen cast them as "a vile set of rogues."[83] In 1829, religious men spoke of the English canal people's "dark and benighted state," their "careless and dissolute practices," and their "decided wickedness."[84] A constable in 1841 accused boatpeople of "a good deal of petty Pilfering; cutting Grass, stealing Turnips, poaching and breaking into Hen Roosts and Things of that Kind."[85] According to a canal and coach owner in 1839, all boatmen were, "to a greater or less extent, poachers."[86]

Police commissioners during the 1830s and 1840s suspected that canal people were intricately involved in the country's smuggling networks. A commissioner in Liverpool in 1839 discovered that not far from each inland waterway leading to the port were illicit depots for the reception of stolen goods. He testified that the same depredations were carried out "upon every canal."[87]

Figure 3.3
Barge families on the Rochdale Canal, Lancashire, UK, winter 1895.
Source: Harry Hanson, *The Canal Boatmen, 1760–1914* (Manchester: Manchester University Press, 1975).

Harry Hanson, a twentieth-century historian of the English canal people, notes that smuggling was indeed quite prevalent on many of the canal routes, especially the smuggling of alcohol, sugar, and dry goods. Lack of confidence in the boatmen led some carriers to require that a portion of the boat be caged up, to protect the wine, spirits, and furniture. Hanson further notes that "there was a desire among the boating people to be accepted into society, but the rejection by society, real or imagined (usually real), played an important part in creating an inward looking sub-culture, a sense of inferiority, and a sense of abandonment among the boating class." Thus, in Hanson's view, a propensity for theft and dishonesty became a part of the canal boatpeople's reproductive culture. Boatmen usually picked up their criminal ways as boys, and were taught the methods of pilferage by the older boatmen. Lockkeepers participated in boatmen's illegal activities as well, assisting in the smuggling trade and engaging in illegal commercial transactions.[88]

These distinctly negative views held by Britain's officials and middle classes regarding the canal boatpeople were also bound up in the sense that the boatpeople constituted a strange "class apart" or "race apart" within British society. Indeed, while mainstream Victorian perceptions of canal people were somewhat akin to Victorian attitudes toward other groups of geographically mobile, subjugated laborers—such as vagrants, who were mostly lone, unattached migrant workers—the ability of canal people to carry their families with them over the course of their travels often implicated them in a pattern of social reproduction perceived as quite separate from the social reproduction patterns characterizing the rest of British society. Thus an 1841 canal trader testified before the House of Lords that the canal people were "a Class, in short, apart from Society—quite distinct."[89] A reverend of the same year submitted that they were "a class of Men *sui generis*. Their Habits and their whole Lives are detached as it were from those around them. They associate altogether" and harbor unusual "Feelings and Views."[90]

Another reverend submitted,

I should know a boatman wherever I see him, his Features, Dress and everything marks him as a distinct Being in Society. There can be almost no mistaking his Appearance. In my Neighborhood they intermarry almost universally with each other, and hence from Generation to Generation their Manners and Habits are perpetuated.[91]

In Britain, references to the canal people as "a race apart" continued into the twentieth century. Brian Vaughton, a British social historian of the 1960s, wrote two BBC radio programs on the slow decline of England's commercial narrow boats:

By the early 1960s not only were the commercial narrow boats disappearing, but also the complete way of life of a race apart—the boatpeople, and the landsmen who worked on the "Cut." For centuries the boatpeople, independent folk at the best of times, had followed their unique way of life, with their own customs and traditions.[92]

Several years later Stella Davis, an English historian of the Industrial Revolution, similarly described England's nineteenth-century canal workers as "a separate and distinct community—a race apart."[93]

Such phrasing is evoked by Hanson as well, writing about the common perception that the British canal people were "Romany" in origin.[94] Hanson allows that "one particular difference common to these wandering tribes, like gypsies, was the distinctive style of dress which emerged and persisted, in some respects, into the 20th century."[95] In this vein, an 1873 writer describes canal people as "dusky as gypsies and picturesque in their costume."[96] The British and Irish canal people's habit of painting their boats in lush colors provoked, among some, a similar set of interpretative reactions. In contrast to the grays and subdued hues of the Anglican churches in the parishes through which they passed, the narrow boats were typically adorned with bright gold, greens, and reds, as well as "crockery and other camouflaging objects." Further tempting the association with the Roma, it was not unusual for boatpeople to keep on deck a wild bird in a gold cage and a mongrelized greyhound, the latter either to guard the boat or, as one police commissioner opined, as a "decided poaching dog."[97]

Nonetheless, Hanson stresses the likelihood that canal people were generally not Roma but rather Welsh and Irish who migrated to the English interior seeking work during the Napoleonic Wars. This conflict acted as an important impetus for the expansion of Britain's canal population in the early nineteenth century. Coastal shipping became vulnerable to French attack, and land carriage rates became prohibitively expensive because horses were being used for battle. To fill the void left in coastal and overland transport systems, carriers developed the so-called fly-boat, which took its name from the small, nimble sailing craft prevalent in Dutch seaports but was in fact a long, narrow canal boat designed to move a considerable bulk of material quickly and without much draft. Fly-boats were pulled by relays of horses and four men, two of whom would rest in a cabin so that the boat could operate twenty-four hours a day. By the 1830s and 1840s, a complex system of fly-boat services had reached its most developed state, operated by some 40,000 people, with vessels commonly traveling both day and night and major boating centers emerging at otherwise remote inland villages, such as Braunston and Shardlow.[98]

The "peculiar" character of the inland boatpeople—their historical association with migrant or wandering groups, their apparent racial distinctiveness, their moral habits and cramped living quarters—all this eventuated in a series of "welfare" interventions, launched by social reformers between 1877 and 1899, to bring the boatpeople closer to the mainstream of late Victorian society. One historian has suggested that these reforms were "the first instance of legislation which proposed to regulate completely the family lives of able-bodied, self-sufficient, contributing members of society."[99] This may have been the case for England, but elsewhere in the British Empire, for instance along the Ganges River in India, imperial authorities had already begun regulating the lives of inland boatpeople—the "Mallah" caste—by the 1860s. Assa Doron highlights the importance of the Ganges as a shipping channel in nineteenth-century British imperial India and the problems, both real and perceived, that the Ganges boatpeople presented to imperial authorities. Elites during the colonial period relied on the Ganges boatmen for the navigation of brassware, silk, zari manufacture, and furniture down the river waters to the Bay of Bengal. Yet at the same time, these elites, both British and Indian, perceived and cast the boatmen as a special "criminal caste."[100] Much as with the English canal people, the Ganges boatmen were also associated with and blamed for thefts, pilfering, and the geographic spread of cholera.[101] Moreover, during the Sepoy Mutiny, the Ganges boatmen were seen by British officials as distinctly implicated in the spread of the revolt, as well as in the betrayal of the British garrison at Cawnpore, which led to the bloody massacre of colonists at Sati Chaura ghat.[102]

Perturbed by this floating population in the heart of their Indian possessions, British colonial leaders imposed the Rules for the Navigation of the River Ganges in 1867, which required that each boatman provide authorities with information about his caste, patrilineal descent, place of residence, and the length of the river under his pilotage. Imperial authorities became even more apprehensive about the Ganges boatpeople during the 1870s, by which time commercial traffic had been partly diverted to new railway lines and carriage roads. The city of Varanasi (Benares) remained a major river boating center by virtue of being a holy city, whose sacred ghats— steps leading down into the restorative river waters—drew pilgrims from throughout the Ganges River network.[103] In 1871 the Mallah caste of river boatmen was included in the Criminal Tribes Act, which provided phrasing to enforce curbs on the Mallahs' mobility within the Ganges system and to forcibly conscript violators into such labor projects as the laying down of new roads and rail lines.[104] To further buttress administrators' ability to intervene in the inner workings of the complex fluvial transport geography of the Ganges, imperial authorities passed the North India Ferries Act in 1878. The Ferries Act sharply restricted navigation on the river between communities and limited the commercial activities of boatmen by prescribing "appropriate" and legally sanctioned methods for plying boats and carrying cargo. The aim of these regulations was only secondarily economic control; the Rules for Navigation, the Ferries Act, and the inclusion of the Mallahs in the Criminal Tribes Act were primarily exercises in social control, and in this sense they prefigured the English canal legislation to come some years later.[105]

Victorian educators and sanitarians advocated for social reform on the English canals during the 1870s, 1880s, and 1890s. Concern for the moral fiber of the English floating population dated back to the 1820s, when conscientious social reformers had launched their "little schools," "Boat-men's institutes," and "floating churches," as well as periodicals like *Boat-man's Magazine*, for the social benefit of the boatpeople and their families.[106] By the late 1870s, though, such social efforts had taken on a more urgent and aggressively interventionist tenor, advocating the imposition of the Canal Boats Act for the enforced regulation of boatpeople's lives, includ-ing their social and reproductive activities. The extent of these regulations generated something of a controversy among some of Britain's educated classes. One *Times* editor objected, in the June 4, 1877, issue, that

the floating home of the "bargee" is to be invaded. Its privacy is attacked. Its liberty appears doomed to pass away ... with a display of inquisitorial power such as was never before dreamt of by any man conducting his boat through the canals and canalized rivers of England.[107]

The Canal Boats Act's political supporters won the day nonetheless. They were led by a minister named George Smith of Coalville, a zealous and determined man who, according to his biographer, believed he heard the voice of God. His 1875 booklet *Our Canal Population* created a stir among reform-minded Victorian audiences. Smith enumerated various statistics regarding boatpeople's immoral practices.[108] He recommended that chil-dren be prohibited from living on boats, that boat cabins meet certain volu-metric standards, that registration of all floating objects be mandatory, that all boatpeople be certified, that all children be educated in regular English schools, and that boatmen have no more than two children on board with them.[109] Once amended in 1884, to allow for effective enforcement, the Canal Boats Act mostly reflected Smith's recommendations.

In the early years of the Canal Boats Act, Smith and the canal investiga-tors put forth wild claims in press releases regarding their findings on canal people's social behavior: boatmen stealing and murdering infants, boatmen buying and selling their wives.[110] These stories rallied much of the public to their cause. Thus in 1879, the nonconformist reverend Mark Guy Pearse published his novel *Rob Rat: A Story of Barge Life*, which told of the vile living conditions for the children of the canals. In his preface, Pearse asks readers "to what district do these poor water babies belong? Who cares for them? No registrar is troubled at their birth, or attends the later ceremony of their wedding."[111] Describing Britain's canal docks, Pearse points out the "weird terrors of the place."[112]

In his own written stories, George Smith tended to associate the canal people with dangerous anti-imperial political movements, in particular the Fenian Brotherhood, an Irish republican political society of the late nineteenth century. Thus if in 1920 Oldmeadow was suggesting that, in some sort of alternative reality, Britain's canals could have provided the physical arteries of an 1848-style British political revolution, Smith more often associated canal people with radical Irish nationalism—an association that was not entirely surprising since many canal boatpeople had ancestry or associations of one kind or another in Ireland. In his 1881 *Canal Adventures by Moonlight*, Smith disparagingly describes a canal towpath horse as "clad by a boot that would suit a Fenian";[113] elsewhere he describes a run-in with a canal boatman named Absalom, "filled with brains of the kind supplied to Fenians."[114]

During the 1880s, Smith became unhappy with the effectiveness of the Canal Boats Act. As he wrote in a letter to the *Manchester Guardian* of November 29, 1881:

It is a burning shame that a few selfish boatowners and Sabbath breakers should, by intrigue, charlatanism, plottery and Satanism, throw dust into the eyes of the nation, hoodwink Parliament, and stifle legislation in a matter that concerns national life, health, wealth and vigour. I say it, and I say it with all the reason, burning language, logic, and force of conviction I can command, that on and after January 1882, any little boat-child that dies through the apathy of the Government and machinators' wickedness shall be as a little Abel whose blood shall cry from the ground.[115]

By 1883 Smith had turned his reform efforts elsewhere: to the religious education of England's Roma.

Published in 1920, at the tail end of Britain's missing ship canal era, Oldmeadow's *Coggin* is the first work of English fictional literature to explicitly associate canals with the specter of insurgent mobility and working-class revolt. That it does so by projecting such an uprising so far back in time is noteworthy. Oldmeadow might have set his tale of working-class revenge against "the burgesses" of the English towns during, say, the 1910s or 1870s. Perhaps the workers of Bulford-on-Deme were inspired by the sailors of Kronstadt or by the Communards of Paris. Instead, to tell his tale of canalside revolt, Oldmeadow reaches all the way back to the tumult of 1848—to the last time that England's canals were actually still in the process of being modernized and expanded, and so to the last time that a canal town like Bulford-on-Deme could plausibly have been such a lightning rod for the class tensions in its region. Oldmeadow's exercise in historical projection draws together a nostalgic recovery of canal town life, on the eve of England's canal-building age drawing to a close, with a reimagining of the unrealized power associated with a lost canal-oriented segment of England's working class.

An early scene in *Coggin* involves an imaginative young boy, by the name of Teddie, happening upon a half-painted sign outside a canalside warehouse, where a new store is soon to open. The painter has excused himself to The Barge Aground for a midday drink and meal. The letters MAR lead the boy to imagine that what's coming is a MARKET, perhaps meaning a market garden for fruits and vegetables. Instead the painter, on returning, spells out the letters MARINE ƧTORE—the backward S suggesting that the painter is slightly illiterate. At this sight, the boy's imagination runs wild:

As he pondered upon the possibilities of a marine store, Teddie ... grew contemptuous toward market-gardens.... At the marine store there would be, without doubt, coils of rope, kegs of rum, stout sails, mariners' compasses, marlinspikes, cutlasses, and barrels of ship's biscuit ... a brass cannon or two, and it was not entirely beyond belief that [the owner] might be able to sell a Jelly Roger with a skull and cross-bones. As Teddie stared at the deserted towing-path it did his small heart good to think of the coming changes.[116]

It may be the case that, to cite again the pamphleteer of the 1840s, the railroad's "air of assumption and parade" sparked the imaginations of many Englishmen in the nineteenth century. But as this passage from Oldmeadow indicates, the same could most certainly be true of the canals. For Victorian observers, the canals and their boatpeople sparked romantic associations with gypsies and exotic races; fearful associations with disease, smugglers, and criminality; religious associations with moral depravity and Satanism; and political associations with anti-imperial groundswells. For the bourgeoisie of Oldmeadow's novel, a canal portends a coming Chartist or Socialist rebellion. For a young boy in this same novel, it portends maritime adventures and pirates.

And the use of the canals for not only criminal but also politically rebellious activity was not always a mere phantom. The outlook of the young Teddie, who sees the canal itself as an arm of the lawless sea, hints at the geographic scope and organizational contours of such rebellious forces.

Ribbonists, Fenians, and Waterways

To explain the nonrealization of the English Cross ship canal scheme proposed by the 1909 Royal Commission on Canals and Waterways, some historians have pointed to British parliamentary paralysis around the turn of the century. Especially contentious in the halls of legislative power, and consequential for the ambitions of Britain's canal advocates, were the heated debates surrounding Ireland's political place in the British Empire.[117] Yet such historians point to the Irish debates only to imply that Parliament may have been too busy with urgent political matters—concerning not just Ireland but also labor strife and impending war with Germany—to have had time to attend to major infrastructure projects such as the proposed Cross. What these historians have overlooked are the ways in which the "Irish troubles" may have been raising elite concerns regarding the canals themselves. For in Ireland and in parts of Britain and British North America, the forces of anti-imperial sedition and conspiracy often spread along the canal and waterway networks, organized by secret societies such as the Ribbonists and the Fenian Brotherhood.

The Irish tradition of politically seditious secret societies dated back to the late eighteenth century, a time when Irish peasants, attempting to resist the collapse of traditional agrarian farming, formed secret organizations such as the Whiteboys and the Defenders. Rituals, oaths of secrecy, and the exchange of passwords were prominent in such organizations. Catholic religious identity played a prominent role in the societies as well, not least because of discriminatory British laws prohibiting Irish Catholics from possessing arms.[118] Society members' activities included the distribution of politically incendiary anti-British literature and the formation and execution of plots to assassinate hated landlords. Irish secret societies also played a fluid social role in their surrounding communities, acting alternately as unofficial police forces, job-protection organizations, racketeer rings, smuggling and black market networks, and gangs for local male youths.[119]

Ribbonism emerged from this social and political tradition in the 1820s and peaked in activity during the 1840s and 1850s. Its name referred to the green ribbon worn by some Irish dissidents to protest "Orangeism," which supported the unification of English, Welsh, Scottish, and Irish lands under a Protestant monarch in London. Ribbonism differed from previous secret societies in its class and urban character. Whereas previous societies had drawn heavily on the publicans and artisans of the outlying agrarian regions, the ranks of Ribbonism drew heavily from Ireland's rapidly growing urban working class, especially after the Irish potato famine. The society was strongest in the areas least affected by the famine-period peasant disturbances; important Ribbon lodges were in Dublin, Belfast, Londonderry, and the city of Limerick.[120] Moreover, Ribbonism spread to the cities on the west-facing coast of Britain more readily than it did to the remote agricultural regions of western Ireland. Ribbon lodges existed in Glasgow, Liverpool, and Manchester, all gateways for Irish migrant labor into Britain.[121] These British outposts of Ribbonism were important enough in the movement that the Dublin Ribbonists' administrative papers, apprehended by police during a raid, referred to Liverpool as the society's "committee in the North," entitled to send two delegates to the most important society meetings, in Dublin.[122]

Ribbonism's lodges, referred to as "markets" in the Ribbonist secret lingo, were perennially involved in smuggling. On the Liverpool and Dublin waterfronts, Ribbonism functioned as a form of primitive trade unionism, or a protection racket, seeking to corner niche labor markets by threats and violence to outsiders and by conducting beatings and even murders to ensure conformity among rank-and-file members.[123] Newly arrived migrants not familiar with "the goods"—the latest secret Ribbon grips and passwords—often found themselves unable to find work, and subjected to physical intimidation if they tried.[124] The lodges also provided tramping benefits to itinerant migrant workers, or "spalpeeners," as they were known in Irish Gaelic.[125] This sanctuary role was at times extended to republican radicals fleeing British authorities.

Ribbonism enjoyed some sympathy and admiration from the organized Chartists, Socialists, and Communists of the industrialized regions of Britain and Europe. Yet some politically radical commentators, among them Karl Marx, were skeptical of the extent of Ribbonism's political and historical importance, finding the society to be much too decentralized and wanting any sort of overarching program of action.[126] One recent historian, M. R. Beames, has attributed this decentralized character of the society to its organizers, who were recruited from the ranks of Ireland's lower classes and saw themselves simply as distributing incendiary literature and launching small-scale disruptive acts to keep the populace "prepared" for a moment when the Irish *upper* classes would summon the nation to arms.[127] In this sense, the Ribbonists contrasted with what Eric Hobsbawm has called the early nineteenth century's "artisan-mason" family of politically radical brotherhoods, which shared with Ribbonism an affinity for symbolism, signs, passwords, and rituals but also typically perceived their role as that of an elite of carefully selected members with a strategy of "imposing the revolution on an inert but grateful mass."[128] The Ribbonists more closely resembled Hobsbawm's "social bandit" archetype but had a greater urban presence than many of the politically radical bandits treated in Hobsbawm's relevant studies, who are mostly rural rebels.[129]

Lacking a political elite, the Ribbonist organization was nonetheless able to administer such complex logistical tasks as the mass diffusion of radical literature, communication between lodges, espionage on the Irish Sea, and the trade of black market cargo.[130] Ribbonist logistical advantages hinged on the fact that, within and around the cities, it was primarily the carrying classes, and in particular the canallers, who provided the core membership, with little if any membership coming from the destitute urban slum dwellers or the manufacturing trades.[131] So important were the boating people for the life of the organization that in Ireland, the geography of Ribbonism approximated the geographic courses of the Royal and Grand Canals.[132] In this sense, the "elites" Ribbonism drew on were those people skilled in the art of clandestine or illicit transport.

The centrality of the canal boatmen for the operation of Ribbonism is further underscored by the testimony of one Edward Kennedy, a witness at the 1840 trial of the Ribbonist leader Richard Jones. Kennedy, though turned informant, was once a Ribbonist himself, and in the courtroom proceedings he explains how he

was a boatman on the canal. It was on the canal I traveled then. I distributed these papers to the Masters on the way. The Masters would know the day I would be going down…. I don't know how strong the Ribbon Society would muster. I was in the habit of meeting them in Dublin. There were many lodge rooms in Dublin. There were not twenty; nor ten; about five, perhaps. About thirty would meet at a time. I don't know how many we mustered at the Cobourg Gardens.[133]

In the text of the courtroom documents, a lawyer asks this informant whether there were as many as "five hundred Ribbonmen there," at the place called Cobourg Gardens, to which Kennedy replies:

I don't know. I went with all the Canal Harbour men. They were not all Ribbonmen. There were, I suppose, twenty of them. There was a great shindy. I took no part in it. I liked the fun very well…. After the meeting at the Cobourg Gardens we met at M'Kenna's on the Quay. That was a lodge-room. Some of the coal-porters were there. Of course they would not be there if they were not Ribbonmen.[134]

The lawyers show special interest in the Ribbon papers Kennedy distributed along the canals, asking him to explain whether they were "printed or written," repeatedly requiring him to explain the meaning of passwords like "the market" and "the winter is approaching" and to explain who authored these passwords and how their meaning was transmitted. Kennedy is evasive in translating the passwords quoted to him, but he clarifies that they were changed every three months, that he "did frequently write some of the passwords" himself, under the direction of leaders like Jones, and that he gave the "regulations" along the canal as he went, often paying out money.[135]

The importance of the canals to the Ribbonists' organization is further suggested by Kennedy's response to cross-examination from Richard Jones's lawyer, Francis McDonagh, who attempts to discredit Kennedy by drawing the court's attention to the witness's past involvement in a canalside murder. Their exchange indicates not just the violence and rough culture of the Irish canals but also the Ribbonist members' feelings of group "ownership" over the canals as a physical space:

Kennedy: Rooney lived in Dublin, he was killed some place about the county Kildare. It was on the bank of the Canal.... I was not the man that planned the murder of Rooney. I did know well the man was going to beat him. I begged of him not to do much to him.

McDonagh: To take him tenderly?

Kennedy: Just so. I don't know what weapon he was to use. I did not see any weapon in his hand. I told him not to do much to him, but just to send him back. He was to get a beating for being in the boat.

McDonagh: —For what right had he to be there?

Kennedy: Exactly so. I don't say he was a colt. He was not wanted there. He was not injured by me.... He would have got more if I did not take him into my boat.[136]

Later in the trial, another of Jones's lawyers, a man named Brewster, uses the Kennedy testimony to shift blame for the Ribbonist crimes away from Jones, the society leader, and toward Kennedy and his ilk of boatmen:

Gentlemen, I am afraid there is or was a society that did exist in this country called the Ribbon Society; and I fear it existed in this town: and I do not entertain a doubt that the witness, Kennedy, the boatman, the murderer ... was connected with that Society. I do not entertain a doubt, that when arrested and likely to be punished, he thought it would be very convenient to screen his own guilty associates, and put the punishment they had earned upon others; and for that purpose he may have endeavored to fix upon Dardis [another Ribbon leader] and Jones the guilt belonging to himself and the boatmen upon his Canal. It is plain that Canal Harbour is the great centre from which the Ribbon system is derived.[137]

Despite the scene of this trial and the subsequent 1842 raids of many suspected Ribbon lodges in Ireland, and several lodges in England and Scotland, the movement's capacity for anti-imperialist agitation, organized crime, and overall troublemaking continued unabated throughout the middle part of the nineteenth century, provoking a further parliamentary investigation in 1859.[138] Equally unwavering was the secret society's hold on the Irish canals. By 1871 the movement was in decline in England, but perseverant in the Irish counties along the Grand and Royal Canals. In this year, the Lord Lieutenant of Ireland, John Wodehouse, holding the aristocratic title of Earl of Kimberley, complained before a session of Parliament of the "Ribandist conspiracy" in the county of Westmeath:

It is a remarkable fact that, whereas this conspiracy was formerly directed almost entirely to questions arising out of the tenure of land, it has of late years begun to extend itself to all the relations of society, interfering with railways, canals, and almost every transaction between man and man.[139]

Kimberley noted that though these conspirators were interfering with the operation of the Midland Great Western Railway, they were for the most part not themselves railwaymen:

In the case of the railway company, the directors resisted the conspiracy; but in the case of the canal, the directors entirely yielded to it, and the canal is, in point of fact, in the hands of the conspirators.[140]

Kimberley also highlighted an incident in which a railroad stationmaster at Trim was shot and killed by canal Ribbonists as punishment for having fired several porters (who were probably society members). Ever since, the stationmaster at Trim "has lived in a bullet-proof house, guarded day and night by policemen; and this in a town for hundreds of years within the English pale, in the centre of one of the richest counties in Ireland."[141] Invoking the testimony of a railway operator recently charged with the task of managing a canal parallel to the line, Kimberley traced the canal conspiracy—at least "twenty years old"—to canal workers' "having been allowed for years to have all the appointments to themselves. As soon as a man dies or goes away the company is told whom to put into his place. These people act as a sort of police, and protect the place." Kimberley offered much praise to the railways for having successfully suppressed this sort of behavior "and in a district where there is so much difficulty in obtaining assistance in putting the law in force."[142] Whether Ribbonism could all by itself have alarmed British elites enough to discourage them from proceeding with important canal plans is debatable. Yet, much as the *Coggin* character Teddie imagines the inland canals as an arm of the adventuresome high seas and waterways beyond, the use of waterways for anti-imperial movements extended beyond the canals of Ireland and Britain—for example, to the border waterways between Canada and the United States. Here, members of an Irish secret society called the Fenian Brotherhood, whose strategic agenda and geopolitical aims were much more explicit and ambitious than those of the Ribbon Society, ran raids from the United States into Canada in 1866 and 1871 with the aim of pressuring Britain to liberate Ireland from the empire. The Fenians' most spectacular raid, in 1866, entailed the secret movement of thousands of radical guerrilla fighters by canal barge in the region around Niagara Falls. Although in the chambers of imperial power the Fenian raids did not produce a scene comparable to the Lord Lieutenant of Ireland's speech before Parliament condemning the Ribbonist conspiracy on the Irish canals, some discussion of the political and logistical precedent the Fenian raids set is still essential for more fully placing the subsequent pattern of waterway divestment throughout the British Empire in its more complete social and political context.

The Fenian raids are well known among historians of Canada, of the United States during the Civil War period, and of the transatlantic Irish in the nineteenth century. The raids have been written about as, alternatively, a watershed moment in the making of Canadian national political consciousness,[143] or an important episode in the Irish independence movement,[144] or an example of American "filibusters," nonauthorized invasions by American militias into smaller countries, the most famous such incident being William Walker's brief seizure of power in Nicaragua.[145] The Fenian raids have not been analyzed in terms of their geographic features—an examination of which reveals the importance, for the Fenians, of the boat and waterway system running between the United States and Canada, from the Bay of Fundy on the Atlantic seaboard to the Great Lakes of the continent's interior. A geographic analysis of the raids also illuminates the special difficulties and frustrations the raiders encountered in attempting to commandeer the border country's railway grid.

Fenian commanders were trained combatants, veterans of the American Civil War from both the Union and Confederate sides. The Fenians also drew their ranks from the extensive Irish secret societies prevalent in North America in the mid-nineteenth century: Whiteboys, Ribbonmen, Hibernians, Shamrockers. Such gangs and organizations were, much like their British Isles counterparts, prevalent on urban waterfronts and along canals.[146] The communication networks and forms of political consciousness fostered by these societies' geographic extent provided one important foundation for North American organized labor activism, as many Irish American workers in the mid-nineteenth century, especially in urban trades, began substituting strikes for violence.[147]

Irish American secret societies also agitated for Irish independence and, it was hoped, for the breakup of the British Empire. This was the grand agenda dominating the council of the Fenian Brotherhood at Cincinnati in 1865. Of special interest to the Fenian leaders was the fraught relationship following the Civil War between the United States and Britain. Political leaders in Washington were demanding reparations for the British role in supporting the Confederacy, in particular Britain's having provided Confederate rebels with a Canadian land base from which to launch wartime raids into Vermont. In the meantime, British leaders were troubled by what they saw as American support for Irish terrorist networks.[148] The Fenian leaders hoped to manipulate and marshal these sorts of geopolitical tensions and fears to their cause, that of creating an Irish republic with a constitution modeled on that of the United States.[149] As these leaders declared at their 1865 conference at Cincinnati, such a republic's temporary home was to be in some piece of Canadian territory, likely Quebec's Eastern Townships, with the city of Sherbrooke acting as the capital of the "Irish Republic in exile."[150] Then, once a humbled Britain had liberated Ireland from the empire, this "young giant Republic" (as Marx described the Fenian specter) would eschew Canadian in favor of Irish soil.[151]

To secure this temporary Canadian home front, Fenian commanders spent 1865 plotting coordinated raids on southern Ontario, Quebec, and New Brunswick to cut important communication lines and paralyze the movements of the British army and any Canadian volunteer army the British could muster. The Fenian commanders purchased boats, munitions, and tens of thousands of guns. They expected their army to be 80,000 strong—"a greater number than were ever before mustered to the conquest of the Canadian possession." Out of this massive army, some men were to "equip a navy on the Lakes Huron, Erie and Ontario," others to move down the Saint Lawrence on Kingston, and others to converge on and lay siege to Montreal; while in the Maritimes, "isolated expeditions from the rendezvous at Saint Andrews" were to "reduce Saint John and Halifax, these [ports] furnishing depots for privateers and ocean men-of-war to intercept British transports and effectually close the Saint Lawrence."[152]

Thus, the use and control of waterways was a central strategic element of the invasion plan. In part this strategy reflected the geographic realities of the U.S.-Canadian border. Yet it may also have reflected a more deep-rooted Fenian familiarity with clandestine movement by boats and waterways. In 1865, Fenian leaders in Ireland attempted to smuggle a cargo of rifles from America to a secluded bay on the western coast of Ireland on board the small ship *Erin's Hope*, though they were spotted on arrival by a British man-of-war.[153] During a citywide police raid and suspension of habeas corpus in Dublin in 1866, Fenian members darted to the waterfront steamers, where many were able to dodge the police.[154] In 1867, two Irish American Fenians involved in an abortive attempt to seize arms at a castle in Cheshire, England, attempted their escape by way of a Dublin-bound collier brig called the *New Draper*, knowing that the passenger steamers would be under surveillance. When the two Fenians arrived in Dublin they were smuggled onto an oyster boat but spotted by harbor police. An elaborate chase ensued involving the oyster boat, a canal boat, a ferry, and a collier, but ultimately the two insurgents were apprehended.[155]

In the international boundary region around Niagara Falls, where the Erie Canal meets Lake Erie and the Niagara River, the Fenian filibustering scheme was attractive to many canallers. During the mid-nineteenth century, the upstate New York canal system was well populated with experienced people-smugglers, many of whom had assisted in the secret transport of escaped slaves northward during the 1840s and 1850s.[156] The importance of the canal watercraft for the Fenians was especially pronounced in the largest Fenian filibuster, into southern Ontario in June 1866. These raids were led by the Fenian commander John O'Neill. In the days leading up to the Niagara raid, thousands of "mysterious strangers" were seen gathering in Buffalo and other American towns along the frontier. These were armed Fenian outfits from Tennessee, Kentucky, Indiana, and Buffalo itself.[157] Some records indicate that among these Fenian ranks were also five hundred Mohawk Indians and a company of one hundred black veterans of the Union Army.[158] A writer of that year recalls that

in Buffalo there were more resident Fenians than in any of the border cities; and, the immense amount of shipping in the harbours of Buffalo and Black Rock, rendered it easy for the Fenians to procure the means of effecting a crossing, while the enormous amount of trade which is continually going on there, the active movements, hither and thither, of numberless canal boats, tugs, schooners, and steamers, employed on legitimate business, rendered it almost impossible for the United States authorities to search out and discover which particular boat, or set of boats, was engaged to carry over the Fenians.[159]

The canal barges were waiting for the filibusterers and their munitions at Pratt's Iron Furnace Dock in Buffalo. Five hundred soldiers at a time, along with their supplies, were able to fit onto the barges. At about midnight of June 1, the tugs, barges in tow, quietly slipped out of Buffalo Creek, avoiding the vigilance of American authorities, and set off for the Lower Ferry Dock on the Canadian shore, about a mile from the British-manned Fort Erie on the Niagara River. Some years later, a field commander described the scene:

Just as the boats struck the shore, the color-bearers of Col. Owen Starr's 17th Kentucky Regiment sprang on to Canadian soil and unfurled their Irish flags amid terrific cheering by the Fenian troops.... There were no Canadian troops whatever within 25 miles of Fort Erie.... The war material was quickly unloaded from the canal boats.... The total number of troops that came over by the first boats was stated to be 1,340.[160]

These initial waves of canal-boat Fenians easily took Fort Erie, and some three hundred of them proceeded inland to ambush Canadian volunteer forces at a village called Ridgeway. General O'Neill used a messenger in a rowboat to relay his instructions to each new wave of Fenian barges from Buffalo. The major gunship in the area, the USS *Michigan*, had been sabotaged by Fenians the previous day and was not able to disrupt these shipments. During the afternoon, though, the *Michigan* was finally repaired and approached the Niagara River. O'Neill attempted to communicate to his forces in Buffalo to stay clear of the river, but canal barges carrying hundreds of Fenian reinforcements were already halfway over the water, and were intercepted by the gunship. O'Neill and his principal officers were apprehended as well, and taken on board the *Michigan*, while the rank and file were left huddled on the canal boats for the night.[161]

Figure 3.4
The lake gunboat USS *Michigan* was able to intercept an Erie Canal barge carrying Fenian filibusterers, 1866.
Source: Frank Leslie's Illustrated Newspaper, June 23, 1866, 211.

In the meantime, the two thousand or so Fenians on the Canadian side of the Niagara, now sounding a retreat, were able to elude both British Canadian and American authorities by taking their canal boats back across the water under cover of night.[162] Some officials claimed that many of these "Leanders" drowned in the crossing, others that they had reached the city on the American shore and subsequently disappeared. Either way, by the time British and Canadian reinforcements arrived in the vicinity of Fort Erie, there were few Fenians left for them to capture or interrogate, and thus every reason for them to imagine that tens of thousands of energized Fenian soldiers still lurked in Buffalo and beyond. Though a failure in terms of its stated field aims of literally establishing an "Irish Republic in exile," the waterborne raid on the Niagara region succeeded in generating a deep sense of alarm among British authorities regarding their capacity to hold their North American borders against the forces of antiimperial revolt.

By contrast, Fenian attempts to commandeer the railroad system along the international border in this region were far less successful in generating such trepidation. To access the scenes of invasion into southern Ontario and Quebec, thousands of Fenian soldiers attempted to use the railroads of New York State and Vermont. The American general George Meade, determined to stop these movements, forbade the American railways and other transportation companies from carrying war materials to the frontier.[163] These orders were rigidly complied with by the companies, and seizures of arms and ammunition were made at train depots throughout the North Country of New York and in the Lake Champlain region of Vermont. At Watertown, New York, a carload of Fenians attempted to resist the seizure of their train, but the local U.S. Marshals were obdurate and sidetracked the cars containing the munitions. The momentarily subdued Fenian soldiers remained in Watertown throughout the afternoon, plotting the recapture of their supplies. The marshals left only a few guards to protect the cars, and that evening the determined Fenians not only recaptured their munitions but also commandeered the evening express train. Yet this filibuster of "train riders" was short-lived: their train was apprehended at De Kalb Junction, well before it could reach the frontier.[164]

Several Fenian raids also occurred at this time in the region around Passamaquoddy Bay, where the St. Croix River, separating Maine from New Brunswick, meets the Bay of Fundy. The many coves, inlets, and islands of this area had been popular among border smugglers, pirates, and privateers since the seventeenth century.[165] The Fenian raiders in this region intended to occupy one such island, Campobello, and turn it into a base from which to run sabotage operations targeting British ships coming in and out of Saint John, the empire's major shipbuilding center in North America.[166]

As in the Niagara theater of the revolt, the raids around Passamaquoddy did not achieve their aims in the field, but still made for quite a spectacle. Telegraphs from dispatchers in Eastport and other towns during the Passamaquoddy raids capture a local sense that the waterways of the region had been commandeered by lawless seamen and privateers. One dispatch of April 22 reported that some fifty Fenians had hired a Portland schooner by the name of *Two Friends* and were heading for Lubec. The ship's captain refused to take them when he noticed their weapons, but they turned the guns on him and forced him onward. They then continued on in their commandeered vessel into New Brunswick waters.[167] An April 25 dispatch quoted a New Brunswick politician as referring to the Bay of Fundy as "alive with ships of war." A May 2 telegraph transmitted a story about a schooner "said to be a Fenian Privateer," which

was boarded by the Custom House officers this morning. A large number of armed men were on board, and she had Fenian arms on board. A Custom House officer just arrived from Lubec reports that armed Fenians left there in a small fishing vessel this morning, but were put on board a large schooner back of Grand Manan, N. B. The United States steamer *Winooski* has just left the harbour in pursuit of the Fenian privateer.[168]

The "back of Grand Manan," which refers to this island's western coast, may have played a special role in the Fenian aims on the region. With its high, forbidding cliffs and secluded coves, the west side of Grand Manan had been a popular spot among smugglers and pirates for several centuries.[169]

This association between unauthorized filibusters and the specter of waterborne smugglers, privateers, and rebellious boatmen is not local to the Fenian raids on the Niagara frontier and at Passamaquoddy Bay. The etymology of the word "filibuster," which entered the English lexicon as early as the seventeenth century, captures the association quite fully. The ultimate source of "filibuster" is the Dutch *vrijbuiter*, meaning "freebooter"; yet the word first appeared in English as *flibutor*, meaning "fly-boatman" (that is, a harbor boatman or a canaller). Indeed, the *Oxford English Dictionary* allows that there is some etymological ambiguity here, and opines that the English "filibuster" may have a dual origin, or that *vrijbuiter* and *flibutor* may share a common etymological source.[170] Thus the current American use of "filibuster" to refer to an unauthorized or obstructionist political act only partly captures the full richness of the term. This modern usage indicates that a filibusterer is, in a loose sense, a "freebooter," one who engages in a rebellious act against some regime of power. Yet the word's etymological relationship to "fly-boater" suggests that this *political* relation, between rebel and regime, can have a *physical* dimension as well: one in which floating objects are engaged in clandestine movements across physical space, much like the Fenians' movements by canal barge across the Niagara and by schooner across the Passamaquoddy water channels.

For British and American authorities, the lesson to be drawn from the raids was that, during this nonauthorized attempt by bands of veterans to overturn the geopolitics of the North Atlantic, railway infrastructure had proven easy to police and secure, while the waterways had not. This upshot by itself was perhaps not enough to mobilize elite support away from important canal projects in the region and toward the railroads. Yet an incident some years later at the southern frontier of the newborn Canadian province of Manitoba indicated that British officials perceived the unique role the border waterways had played in facilitating the raids, and that such experiences were giving officials considerable pause as they weighed the potential benefits of canal schemes in the Canadian Midwest.

The incident in question was the last Fenian raid, launched from Pembina, North Dakota, into the Red River Valley of Manitoba, in 1871, a short-lived filibuster whose unmet object was to foment a rebellious alliance between the Fenian movement and the Canadian Midwest's French-speaking Métis population against Anglo-Canadian hegemony in the region.[171] Boats and waterways were not involved in this raid; the Fenians had to trek over land through the wilderness of the upper Midwest, from Saint Paul to Pembina, and from Pembina into Manitoba. And yet, several years later, a British International Boundary Commission charged with the task of surveying the complex lake geography separating Minnesota and North Dakota from Canada cited the 1871 Pembina raid, and by extension the more serious Fenian raids of 1866, as a reason not to build an all-Canadian canal in the region that would link the Red River of the North and Lake Winnipeg to the Great Lakes.

Some historians contend that the formation of this British International Boundary Commission—best known, perhaps, for having postmarked the 49th Parallel—was itself a reaction to the Fenian raids, in particular to the events in Pembina.[172] Another historian has deepened this analysis, suggesting that it was not only the Pembina raid but the very geography of the upper Midwest, where the natural and historical transport corridors ran north to south, that urged British authorities to "sever," physically, the Canadian Midwest from the U.S. Midwest, thus narrowing the possibility of secession of the Canadian Midwest or annexation to the United States. Such physical partition depended not only on a long and visible line of postmarks but also on a railroad geography specially contrived to link the Canadian Midwest with Ontario rather than with the American watersheds to the south. Indeed, both Canadian westward expansionists and American annexationists believed that if western Canadian produce was routed south toward Saint Paul and Chicago rather than east on an arduous journey across the Canadian Shield on the north shore of Lake Superior, "the Métis country of Assiniboia and the whole Canadian west would fall to the United States."[173]

Thus the members of the British International Boundary Commission saw as their undertaking not only the marking of the international border but also the surveying of the possible transportation geographies best suited to assert British imperial authority within the region's vast expanse. One surveyor, an Englishman by the name of Samuel Anderson, was asked by the head commissioner to report on the feasibility of digging a canal from the Lake of the Woods whose purpose would be to establish an all-Canadian water route from Lake Superior to Lake Winnipeg and up the Saskatchewan River. Some piecemeal work had been done some years before on the Lake Superior part of this route. After all the surveying and estimating were done, Captain Anderson gave his opinion that a carriage road and railway would be preferable, as a canal would be vulnerable to further hostilities with the United States.[174]

Another Boundary Commission official, D. R. Cameron, supported the idea of digging the Lake of the Woods canal, but voiced his concerns that were such a canal to be built, the so-called "Northwest Angle," a peculiar Lake of the Woods peninsula that is American-held but not territorially contiguous with the rest of the United States, would become a refuge for Canadian bandits, fugitives, and smugglers. For this reason, Cameron, like Anderson, recommended the routing of midwestern freight onto a railroad path rather than a canal.[175]

This perception of railways, and not canals, as a transportation tool well suited for containing and suppressing the smuggling trade along the international border in the North American upper Midwest continued into the twentieth century. In 1909, the writer Elliot Flower published an article in *Pearson's Magazine* on opium smuggling between Canada and the United States. Looking at the border between Manitoba and Minnesota, Flower described smugglers' attempts at using the train cars: "There are many places on a car where such articles can be concealed, and no general inspection takes account of all these. There is the buffet-compartment, the linen-closet, the cubby-holes reserved for various minor details of the equipment, the boxes under the seats, etc." Yet, as Flower explained, for the smugglers, these apparent physical advantages were in fact physical constraints:

This smuggling by the Pullman car route is the prosaic end of the business. It is also the retail department. You can't take very much that way, and it lacks the excitement of smuggling by water. The thrill and adventure are ... found at Detroit, Portland, Seattle, and San Francisco.[176]

At least in part, the decision on the part of leaders in London and Ottawa not to finance any water transportation routes in the Canadian Midwest— thus sacrificing British industrialists' access to cheaply transported North American grain—must be understood as a reaction to such "thrills and adventures" elsewhere in the British North Atlantic domain.

Dempingen

The Dutch word *demping* harbors a curious dual meaning. In the context of town and city planning, the word refers to water infill.[177] Thus, many of the towns and cities in Holland dating from the Middle Ages or Renaissance have at least one modern avenue, cutting across the old town center, with the adjective *gedempte,* "filled in," at the beginning of its name.[178] Gedempte Oude Gracht (filled-in old canal) in Haarlem, Gedempte Burgwal (filled-in moat) in Den Haag, Gedempte Nieuwesloot (filled-in new sluiceway) in Alkmaar—these street names all refer to one-time water transportation and urban hydraulic facilities that state and municipal authorities eventually filled in. By contrast, in the context of social history, especially the history of class conflict, *demping* refers to the suppression or subduing of a riot or social revolt.[179] This was as much the case in the nineteenth and early twentieth centuries, when most of the Dutch canal and waterfront *dempingen* took place, as it is today. For instance, a 1901 English-Dutch dictionary offered as one viable translation of the verb *dempen:* "to smoother, to quell (a riot)."[180] An 1878 Dutch newspaper advertised a public lecture titled "Betreffende de onlangs gedempte opstand in Japan" (Regarding the recently suppressed rebellion in Japan),[181] a 1905 Dutch encyclopedia referred to the *dempen der revolutie* in France in 1848,[182] and so on.

Nor, during the nineteenth century, was this connotative link between hydraulic movements and people's movements lost on some skilled Dutch writers. One 1886 writer for the Amsterdam-based literary periodical *De Nieuwe Gids* likened a bloody working-class uprising the previous summer— the so-called Eel Riot, sparked, as it happens, by administrative attempts to regulate working-class Amsterdammers' use of the urban canals—to a "fiery sea swollen high with the winds of socialism." The writer's subsequent criticism of the failed efforts by police to execute an effective *dempen van dit oproer* (suppression of this riot) employs the double meaning of the word, reinforcing the analogy of raging seas to uproarious masses, and in turn of water infills to political suppression.[183]

Despite the connotative association captured so suggestively by this word in Dutch, the perceived association between canal people and subversive mobility does not show up in Dutch records of the late nineteenth and early twentieth centuries to the same extent that it does in records from Britain. The dual meaning of *demping* may be simple coincidence; it may also stem from social associations dating from a much earlier time. Canal smuggling does seem to have been a major political concern among Dutch officials during the 1860s because of a widespread perception that canal boat pilots were smuggling tainted cattle during a rinderpest epidemic.[184] But after this point, we do not see anything comparable to the British fear of a "canaller" social nuisance or political menace.

The special intensity of Britain's canal anxieties in the late nineteenth and early twentieth centuries—anxieties intense enough to be quite legible in an archival and geographic record—seems to have hinged on the British establishment's distinctive set of fears regarding the possibility of an Irish working-class revolt that, once under way, could conceivably spread across the Irish Sea to draw in discontented elements in England itself. None of the other industrial powers during this period was characterized by a set of geopolitical anxieties quite analogous to the British Empire's "Ireland" problem. The colonies of the Netherlands and Germany were continents away. France, like Britain, had its "inner-ring" colonies: Algeria and eventually Tunisia. But, unlike Ireland, Algeria and Tunisia had hardly any canals or canalling people. The mobile, freight-carrying inland peoples of the Maghreb territories were mostly camel drivers, who—whatever the extent of their guerrilla capabilities in the North African desert—were unlikely to be perceived by anyone in France as capable of "drifting" into the French interior. By contrast, it must have been quite easy to envision the canal people of Ireland proceeding into and up the British canals, where Ribbonist lodges existed and where much of the boating population was already understood as, in some imprecise sense, partially Celtic and Romany. This geopolitical distinction might go some way toward explaining why it is so markedly in the British historical record that we find, intermixed with a half century of dismissed ship canal schemes, zealous reformist initiatives accusing canal boatpeople of vice, criminality, and sedition.

In their 2001 study *The Many-Headed Hydra*, Peter Linebaugh and Marcus Rediker argue that, during the seventeenth, eighteenth, and early nineteenth centuries, the waterborne linkages along and across the North Atlantic Basin enabled a tremendous amount of clandestine movement of weapons, papers, and people among politically rebellious groups, rendering difficult efforts by British imperial authorities to keep separate and localized different elements of a radicalizing North Atlantic workforce. As Linebaugh and Rediker argue, this Atlantic workforce, characterized by the shared experiences of exploitation, violence, freedom, and revolt along the North Atlantic waterways, terrified imperial elites and precipitated a gnawing ruling-class desire either to annihilate the "many-headed hydra" of resistance movements or to clap this hydra in chains.

The opening of the Suez Canal in 1869 demonstrated the viability and effectiveness of a new technology, the ship canal, for extending maritime transport across land. Many advanced industrial nations eagerly embraced this technology. Britain, for the most part, did not. Decisions by British political and financial elites not to invest in new canal construction after 1869—and instead to invest in the novel railway networks, whose usefulness for a maritime empire was limited—must be understood in the context of this deeper social historical background. Linebaugh and Rediker end their analysis in the early nineteenth century, and leave open the question of whether their "hydra" survived the onset of industrialization that was to come mid-century. I would suggest that, at the very least, the *fear* of this hydra, of mobile, marginalized, potentially co-conspiratorial groups, did indeed continue into the nineteenth and even twentieth centuries. And one of the manifestations of this fear was a strong negative bias against canal boat workers.

For many members of the British ruling and middle classes of the late nineteenth century, canal boat workers were associated with immorality, with racial otherness, with ungovernable "wandering" groups, with the spread of disease, with smuggling, pilfering, piracy, and political subversion. In part, such perceptions were quite irrational. There is little evidence suggesting that canal boat workers were especially prone to radicalism or subversion compared with workers in other British trades; moreover, in actuality, during this period workers in such industries as rail transport and mining seem to have had a far greater inclination toward oppositional politics. But such fears also stemmed from the very real unique physical capacities of canal boat workers: the capacity for clandestine movement of people and cargo; the capacity (in the case of the fly-boats and canal barges) to raise a family on board one's own means of conveyance, and so to exert some degree of control over social reproduction. Canal boat workers rendered less tenable a host of politically functional distinctions, in particular between Irish and English labor. They also enabled the diffusion of conspiratorial anti-imperial papers, weapons, and politics along the Irish canals in ways not feasible along the Irish railroads. Similarly, canallers and canal boats proved instrumental—while the railroad grid did not—for the invasion of thousands of radical republican fighters into British North America during the tumultuous years following the American Civil War.

None of this is to imply that subversive groups did not *sometimes* make use of the railways to move goods and people as needed for their causes (or that the canals themselves somehow "made" revolts happen). It is rather to stress that, in the British North Atlantic, an entire history of social biases and fears, of class and civil warfare, violent anti-imperial agitation and the trafficking of rebels' cargo, must be taken into consideration when seeking to understand more fully why, by the outbreak of World War I, British food markets depended primarily on Eurasian rather than Canadian grain, why Birmingham, Westmeath, and Winnipeg were all landlocked, why the Bay of Fundy was cut off from the fastest shipping route to England, and why, finally, canal boat workers found themselves attached to a declining and soon to be abandoned means of livelihood and conveyance.

4   Chenangoes: The Replanning of Freight Flows in New York City

Another modern geographic outcome that might, at least in some instances, be better accounted for by taking into consideration a preceding history of subversive transport networks, as well as of top-down efforts to frustrate real or imagined versions of those networks, is urban deindustrialization. I use only one city's industrial history as an example. In the first half of the twentieth century, New York City was the largest urban manufacturing hub in the world. From a planning perspective, it was in many ways looked to as a model for other American cities, and for some non-American cities as well. Had a different planning paradigm emerged from New York's industrial heyday, this paradigm might well have influenced twentieth-century urban planning theories and policies on a broader scale. Thus, by crafting an in-depth argument about a single place, I hope to cast new light on directions taken, and not taken, by practitioners of urban transport planning and urban development over the course of the twentieth century.

I begin this chapter by highlighting the impressive variety and expanse, by the middle of the twentieth century, of the Port of New York's urban freight-handling facilities, as well as the dramatic extent to which almost all of these facilities had disappeared by century's end. Certain economic factors tending toward a depression of freight-handling infrastructure were in play during this transition—among them postwar economic restructuring, which to some extent eroded New York's manufacturing base, and rampant speculation among the city's real estate interests, which sought the development of offices and high-end luxury complexes in place of manufacturing and freight infrastructural spaces. However, such economic analyses cannot fully account for the historical constriction of freight-handling activities in New York. The main problem with such analyses is chronological. The history of New York's leaders' opting to divest from local freight infrastructure predates by several decades the appearance of the sorts of market pressures against urban manufacturing that are usually emphasized by economic historians of the city. These market pressures started to cut away at the city's manufacturing base during the 1950s—whereas the divestment from infrastructure began much earlier, with local political and financial elites' rejection of proposals during the 1910s to expand New York's freight transport facilities, and with leadership's embrace of plans during the 1920s to rezone the urban core against manufacturing and related freight-intensive activities.

A better understanding of the shift in the city's economic commitments and related transportation geography begins with this earlier period, the 1910s and 1920s, when urban manufacturing was still economically ascendant. The violent geopolitical and class struggles that took place during this era are especially important for understanding local elites' decision making regarding the city's infrastructure and physical mobility systems. To provide a window into the sorts of concerns and anxieties that were motivating much elite thinking regarding the transportative spaces and divisions of labor in New York during this period, I pay special attention to the 1919 memoirs of the captain of the New York City bomb squad, Thomas Tunney. Tunney painted a picture of a port region rife with subversive watercraft pilots and assembly workers making, moving, and setting off bombs all around the port. These harbor boatpeople and manufacturing workers were, in Tunney's telling, associated with a dizzyingly complex, multinational menace of German saboteurs, anti-capitalist anarchists, Wobblies, Bolshevik agents, and anti-imperialist revolutionaries from Ireland and India. Tunney calls the most worrisome segment of the class of watercraft pilots the harbor's "Chenango" workers—a local slang term that may, I suggest, connote a deeper history of local experience with violent subversion. The imagery Tunney presented, as well as the real set of experiences on which this imagery drew, was indicative of (and perhaps helped solidify) a pressing elite bias against the port region's casually employed, or "floating," freight-handling workforce. And such fears in turn helped shape subsequent urban planning biases around New York City from the 1920s on.

Why Doesn't New York City Have a Subway System for Freight?

Nearly all the freight-handling facilities serving New York City's borough of Manhattan in the middle of the twentieth century were gone by the century's end. The elevated freight railroad line on Manhattan's West Side had been abandoned. The 60th Street freight yard and docks at the base of the Upper West Side palisades were gone too, as was the Lower West Side's St. John's Park Freight Terminal, whose unusual if shabby-looking structure, bridging over Houston Street, was later converted to lofts. Along the waterfront on both the Manhattan and New Jersey sides of the harbor hundreds of car float transfer stations, once the principal means of cargo transport across the Hudson River, were no longer in use. A handful had been preserved as architectural follies for new waterside landscaping schemes, but most had been left to decay into the water, leaving behind the occasional sinking "graveyard" of rusting steel and rotting wood for the twenty-first-century observer to look upon with, perhaps, no small amount of puzzlement.[1] Gone too was the West Side Elevated Highway. First constructed in 1929 as the first urban express highway in the world, with extensions added during the 1930s and 1940s, the West Side Highway had functioned to enable a relatively unhindered flow, underneath its viaduct structure, of boxcars, trucks, and carts moving between the piers lining the West Side waterfront and the many factories, workshops, warehouses, and freight yards of Manhattan's interior.

The cargo-carrying capacity bound up in mid-century Manhattan's various modes of freight transportation was not replaced by any alternative freight-carrying technology; this capacity simply disappeared. To be sure, over the course of the late twentieth century, some parts of the port—mostly the remoter sections of the outer boroughs or the New Jersey side of the harbor—received some improvements in motor truck facilities, in the form of interstate highway expansions and truck terminals. But this was not the case in Manhattan, the most economically active and thickly settled section of the port region. As, over the course of the second half of the twentieth century, Manhattan's considerable concentration of lighter-age piers, car float terminals, and boxcar-handling facilities was abandoned and razed, truckers in Manhattan found themselves confined to the very same urban street grid, much of it laid out in the early nineteenth century, that they had been sharing with pushcarts, automobiles, wagons, carriages, delivery bicycles, and pedestrians since World War I. Proposals to expand Manhattan's truck capacity—for instance, plans by the Port of New York Authority to modernize its Union Motor Truck Terminal on Greenwich Street, or plans by Robert Moses for an express truck route in Midtown—came to nothing.

In part, this history of urban freight facilities' becoming abandoned and erased, and of a local field of possibilities for the movement of freight in the city becoming dramatically constricted, can be understood in relation to New York City's changing economic base following the 1950s. In the first half of the twentieth century until as late as the close of World War II, over one million workers—roughly half the city's workforce—had been engaged in manufacturing and freight-handling labor.[2] During this period,

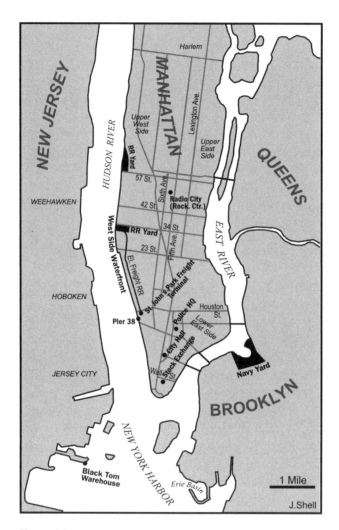

Figure 4.1
Some of the locations discussed in this chapter.
Source: Cartography by J. Shell.

much of the cargo flowing into, out of, and within the city moved through
the urban core's many assembly plants and workshop floors, whether as
raw materials, processed parts in assembly chains involving many different
manufacturers, or completed products ready for market consumption.[3] By
the 1970s, though, New York City had lost nearly a half million of these
manufacturing jobs—nearly 300,000 in Manhattan alone—with tens of
thousands of plants shutting their doors.[4] And this eroding manufacturing

base in turn eliminated many of the local freight-handling facilities' apparent raison d'être.

Some economic historians have sought to explain this decline of manufacturing in New York by looking at wider revolutions in industrial production following World War II. One explanatory approach points to the prevalence, in New York, of craft industries that, except among the most high-end designers, proved to be vulnerable to post-war forms of mass-production and mass-consumption in the United States and Western Europe.[5] Yet, however appropriate this account might be for certain industries in New York, other economic analyses of the demise of manufacturing in the city suggest that such an explanation for the city's late twentieth-century deindustrialization is somewhat misleading. One study headed by Raymond Vernon, for instance, acknowledges that many of New York's manufacturing establishments during the first half of the century were indeed "craft" trades, dependent on a local supply of specialized high-skill labor and perhaps doomed, in an age of mass-consumption, to become obsolete. But this study also contends that many *other* manufacturing establishments in the city functioned instead as "innovative" trades, dependent, like their craft counterparts, on a local high-skill workforce, but also engaged in the development of entirely new technologies and products— transistors, radio parts, electron tubes, medical devices, and so on—that were heavily valued in numerous mass-productive industrial sectors, as well as, in some cases, by military contractors.[6] Indeed, Vernon points to what he calls a product's "life cycle," in which specific industries tended to germinate in the urban core (not only in New York but also in other urban industrial centers), then mature into powerful mass-productive industries. In Vernon's analysis, at this mass-productive stage of a given product's life cycle, industries often located their factories outside urban centers, on cheaper land and near cheaper labor, while at the same time maintaining numerous contracts with, and advancing significant capital sums to, the workshop innovators and experimenters still clustered in the industrial downtowns.[7] As this study observes, this cycle of innovation, migration, and urban replenishment dated at least as far back as 1900.[8] Yet during the decades after World War II, the cycle stopped. Industrialists and military contractors alike ceased to forward capital to innovative assembly trades in the city—a cessation that exogenous market changes, such as post-war revolutions in mass-consumption and mass-production, cannot satisfactorily explain.

In his 1993 *The Assassination of New York*, Robert Fitch offers an alternative explanation for what happened to cargo-intensive manufacturing activities in New York. One of Fitch's key premises is that the health of industries in New York, and especially Manhattan, was vitally dependent on political and financial support from local elites, including local political leaders and, most important, banking, industrial, and real estate leaders. These leaders provided urban manufacturers with various forms of infrastructural investment, including investments for freight transport. Fitch speculates that had these patterns of political and financial backing continued into the late twentieth century, New York manufacturers might have found a place in the post-war urban economic environment.[9]

Yet urban manufacturers did not enjoy any such continued support. Instead, during the middle third of the century, real estate and financial leaders aggressively swung their political and financial backing to infrastructural projects aimed at expanding the city's office-sector economy. To illustrate this shift, Fitch highlights the "family subway" that Standard Oil chairman and New York real estate mogul John D. Rockefeller successfully pressured municipal government to build underneath Sixth Avenue during the 1930s.[10] The main advantage of this project was to drive up the value of Radio City (or Rockefeller Center), the Rockefeller family's four-million-square-foot office super-complex along Sixth Avenue in Midtown Manhattan, as well as nearby office properties held by allied real estate interests. John D. Rockefeller put his son Nelson, then twenty-six, in charge of the Rockefeller Center Subway Committee, the aim of which was to marshal the full force of collusion between real estate capital and government to determine the shape that mid-century transport infrastructure in the area would take. As Nelson Rockefeller wrote in 1933, "The question of transportation … is of vital importance to Rockefeller Center. If a continuous flow of the right sort of people to the Center can be maintained by rapid and convenient transportation, the future of Rockefeller Center would seem to be answered." This matter, he continued, would require "a continuous vigilance—in keeping in touch with civil and governmental projects which affect transportation in a large way."[11] In his analysis, Fitch emphasizes the likelihood that the Rockefeller family's principal lawyer, lobbyist, and retainer, Lester Abberley, was the main instrument through which the family was able to influence transport decision making in City Hall.[12]

What is curious about this history, for our purposes, is the unfailing insistence, on the part of powerful real estate leaders like Rockefeller, on developing Midtown for office use. Throughout the first half of the twentieth century, the Midtown district, especially the zone west of Fifth Avenue—Radio City was constructed on the edge of, but also fully within, this zone—was one of the region's most dynamic manufacturing areas by virtue of its proximity to the Hudson waterfront piers, the 34th Street train yards, and a flexible supply of skilled immigrant workers in the nearby Lower East Side and Tenderloin neighborhoods. However, the major real estate barons in this area opted to develop their properties not as cutting-edge factory facilities requiring new industrial infrastructure (which might include not only new freight delivery facilities but also sound-containing walls, or air-pumping systems to remove noxious fumes from the district core) but rather as cutting-edge office facilities, such as Radio City, requiring new commuter infrastructure.

Nor is the Sixth Avenue Subway the only example of mid-century New York's real estate and financial leaders' using their political influence to push for office-oriented but not manufacturing-oriented infrastructure. Through one of their key vessels of urban advocacy, the Russell Sage Foundation (to say nothing of politically sympathetic media outlets such as the *New York Times*), real estate and financial leaders were able to determine the routing of parkways, the development of parks, the placement of new bridges and tunnels, and the location of slum-clearance and urban rehabilitation projects, all aimed at serving an office-bound urban population. During the 1930s and 1940s, government-appointed public authorities were for the most part obeisant to the wishes of the Russell Sage Foundation in the realization of such modernization schemes. When public authority leaders attempted to ram through projects that powerful Downtown and Midtown real estate interests opposed—such as the Brooklyn-Battery Bridge, proposed by Triborough Bridge Authority chairman Robert Moses in 1939—it was always the real estate moguls, not the government power brokers, who prevailed in the dispute.[13]

The question remains, then, of why factory facilities and industrial infrastructure wound up excluded from financial and real estate elites' vision for the city's economic future—why, that is, inner-city freight infrastructure became an object of divestment, if not in the direct sense of financiers' literally pulling their funds from freight facilities, then certainly in the more indirect but still consequential sense of local leaders' marginalizing industrial infrastructure projects from the vast, capital-intensive modernization schemes that would come to dominate the city's economic and political life mid-century. For it is hardly the case that the office complexes that developers bankrolled with such gusto were the most direct and efficient way to generate a profit. Even though the Sixth Avenue Subway was completed in 1936 (and despite attempts by John D. Rockefeller shortly before the outbreak of war to attract the ascendant Nazi and Italian Fascist parties as tenants), the Radio City office complex didn't turn a profit until 1964.[14] Of course, large office complexes like Radio City were the product of much runaway 1920s office speculation, and it seems quite unlikely that Rockefeller would have proceeded with the development had he known it would take many decades to turn a profit on the behemoth-like building complex. Yet this still begs the question of why New York business and real estate leaders like Rockefeller were so eager from the 1920s on to speculate in office real estate and not in manufacturing real estate.

Fitch, in his analysis, implies that this preference for office development was not altogether economically rational, and hypothesizes that the bias instead may have been social—elites may have perceived manufacturing as incapable of stimulating "the flow of the right sort of people," as Nelson Rockefeller put it, into the heart of New York. Fitch usefully traces this bias back in time, to—in his account—as early as the 1920s, well before any actual trends of urban industrial dissolution had set in.[15] Fitch points, for instance, to the 1929 publication of a prominent multivolume planning study, the Russell Sage Foundation's well-known *Regional Plan of New York and Its Environs*. Backed by several of the city's leading financial institutions, such as Morgan Bank, First National, and eventually the Rockefeller Foundation, the plan proposed the removal of most freight-handling facilities from the city center, the rezoning of the city so as to eliminate most industries from Manhattan, and the construction of a new generation of parks, office buildings, and luxury apartments to line a newly deindustrialized inner-city waterfront.

Of course, Fitch's contention—that a top-down bias against inner-city manufacturing activities actually predated the economic erosion of manufacturing industries in the city by at least several decades—does not disprove the arguments of those economic historians who have maintained that the city's manufacturing economy was doomed by post–World War II global developments in mass-production one way or the other. Nonetheless, this contention does call into question the notion that the causes of New York's deindustrialization are altogether so easy to locate in either space or time. Moreover, such a thesis is also strongly reinforced by a pattern during the early twentieth century—though not a pattern Fitch touches on—of political and financial elites' opting to dismiss heavily-studied engineering proposals to *expand* the city's freight-carrying capacity.

This pattern begins as early as 1908 with the publication of a scheme by William Wilgus, who was the former lead engineer for the New York Central Railroad and architect-engineer of Grand Central Station in Midtown Manhattan, to build a freight subway system for the entire southern half

of Manhattan Island that would run under nearly every street of this heavily industrialized and thickly settled part of the city.[16] As part of his published volume, Wilgus included a cross-sectional diagram demonstrating
how these freight subways were to relate to the streets and buildings around
them. In this diagram, the proposed freight lines run directly beneath the
sidewalks on both sides of a street; track spurs lead to platforms at the basement level of adjacent buildings; from these basement concourses, freight
elevators lead up to what appear in the plate to be sewing floors.[17] It's worth
juxtaposing this street cross-sectional plan, submitted by Wilgus, against an
analogous cross section included in the 1929 *Regional Plan*, which, like the
Wilgus scheme, proposes many subsurface transport tunnels, including six
passenger subway tracks and a tunnel for automobiles, to complement the
street-level transport space, but unlike the Wilgus scheme, proposes nothing to address the problem of freight in the city.[18]

Wilgus's scheme, proposed amid public debates about the future of
the New York Central Railroad's West Side street tracks (known as "Death
Avenue" for the frequency of collisions there between train cars and
carriages), was ignored by local politicians and by the city's major subway-
and el-building companies. A decade later, though, the urban "freight
subway" concept was rearticulated in an ambitious proposal released by
a government-appointed body known as the New York, New Jersey Port
and Harbor Development Commission.[19] Chaired by, among others, rail-
car manufacturer Eugenius Outerbridge, former Republican Party chairman
William Willcox, political counsel and labor arbitration expert Julius Henry
Cohen, and military engineer George Goethals, the Port and Harbor Development Commission proposed a simplified version of Wilgus's decade-old
scheme. Where Wilgus had proposed freight subways beneath almost every
street in southern Manhattan, the Port and Harbor Development Commission's 1920 *Joint Report with Comprehensive Plan and Recommendations* called
for a small-batch freight subway system composed of a just a few main
lines, which, much like a passenger subway system, were to stop at freight-
handling stations every few blocks, rather than at every building.[20] The
principal freight line was to wind its way through the heavily industrialized
Lower West Side of Manhattan and duck under the Hudson River by way
of two tunnels to link up with the large rolling-stock yards of Jersey City.[21]
The *Comprehensive Plan* engineers proposed additional, if lower-priority,
branches as well: a loop branch to access the Wall Street area and the Lower
East Side of Manhattan, a branch to access the 57th Street neighborhood
in Midtown, branches to reach the Upper East Side and Upper West Side of
Manhattan, as well as the South Bronx and Harlem, and finally branches

Figure 4.2
William Wilgus's 1908 proposal for a freight subway system in New York City.
Source: William Wilgus, *Proposed New Railway System for the Transportation and Distribution of Freight by Improved Methods in the City and Port of New York* (New York: Submitted to the Public Service Commission of the First District, by the Amsterdam Corporation, 1908).

Figure 4.3
A street subway scheme in the 1929 *Regional Plan* included six tracks for passenger transport but—unlike the 1908 Wilgus scheme—none for freight.
Source: Regional Plan Association, *Graphic Regional Plan*, vol. 2, *Building of the City* (New York: Russell Sage Foundation, 1931), 395.

in New Jersey to access sections of Jersey City, Bayonne, and Weehawken. A 1922 map by the Merchants Association of New York of manufacturing activity in the city provides a sense of the sorts of cargo these branches might have carried on a day-to-day basis. The Wall Street area was, at that time, the site of many of the city's bookbinding, lithography, and coffee roasting shops; the Lower East Side was a hub of the men's clothing industry; many of the city's more lucrative women's clothing factories had been moving to the 57th Street neighborhood throughout the 1910s and early 1920s; the East River bank of the Upper East Side contained many of the city's large food-processing plants; nestled in Bayonne, Hoboken, Jersey City, and Weehawken were many of the port region's chemical and munitions plants; and the wood trades dominated the manufacturing scene in the South Bronx. As for areas such as the Upper West Side and Harlem, the Merchants Association described these zones simply as "available for industrial development"—signaling that urban manufacturing was still understood, at this time, as ascendant.[22]

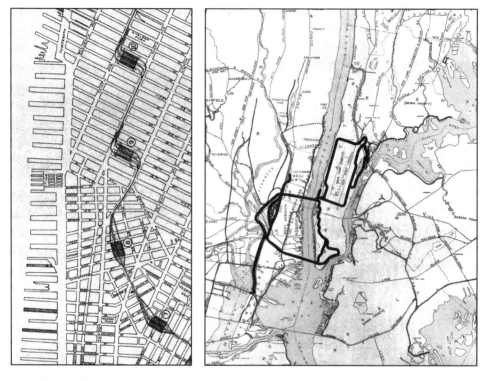

Figure 4.4
Left: The Lower West Side Line of the Port and Harbor Development Commission's proposed freight subway system for New York City. Right: The thick black lines here (enhanced by the author) represent the full extent of the Port and Harbor Development Commission's urban freight subway proposal.
Source: New York, New Jersey Port and Harbor Development Commission, *Joint Report with Comprehensive Plan and Recommendations* (Albany, NY: J. B. Lyon Co., 1920).

The Port and Harbor Development Commission's 1920 proposal was unique in its technological approach. Rejecting the idea that Manhattan's densely packed neighborhoods and industrial areas were suitable for conventional freight trains, the commission authors asserted that a "radical departure from anything found in New York, or in fact in any port," would be necessary to secure for the city "continuous streams in and out of goods of all kinds."[23] The proposed small-batch freight subways were to be partially automated, with switches positioned on each boxcar to direct vehicles to specific stations. At the stations, freight workers were to direct boxcars onto large freight elevators, which would move the cargo upward into cargo terminal buildings enveloped by spiraling vehicle ramps for city trucks and carriages.

Figure 4.5
A prototype of the Port and Harbor Development Commission's proposed freight
delivery stations.
Source: New York, New Jersey Port and Harbor Development Commission, *Joint Report
with Comprehensive Plan and Recommendations* (Albany, NY: J. B. Lyon Co., 1920).

An additional technological innovation detailed in the *Comprehensive
Plan* was the plan's proposed raised automobile "esplanade" to line the Hudson River, on the West Side of Manhattan, replacing the New York Central
Railroad's "Death Avenue" rail tracks and opening up street-level concourse
space for trucks and car float boxcars at the West Side's river piers. As it happened, this esplanade was the only part of the Port and Harbor Development Commission's scheme to wind up being built. An esplanade, or raised
express highway, much like the one proposed in the 1920 plan, opened in
1929 as the West Side Elevated Highway—or, as it was officially known,
the Miller Highway, named for the Manhattan borough president who had
been primarily responsible for the structure's funding and construction.[24]

The *Comprehensive Plan*'s proposal emerged from the leveraging of a
considerable amount political capital, immediately after World War I, on
behalf of the formation of a centralized, coherent administrative authority
for the Port of New York. During the war, military planners and logisticians
had been aghast at the difficulty of moving Europe-bound grain and war
materials from the nation's interior through the port, where jealous and
counterproductive competition between the railroads and inefficient cargo
transfers from train to harbor boat had impeded transatlantic shipping.[25]

Over the objections of political leaders in both New York and New Jersey, as well as of numerous railroad companies fearing for their local freight monopolies, the Interstate Commerce Commission asserted its authority to establish a dual-state body to replan and administer the port, chartering the Port and Harbor Development Commission in 1918.[26] Despite this initial application of political leverage on behalf of the harbor commission's *Comprehensive Plan* scheme, however, all political and financial support for the proposal rapidly eroded over the remainder of the 1920s. The major railroad carriers wanted nothing to do with the plan, nor did the major financiers on Wall Street or any of the city's major real estate interests.[27] In Washington, the level of empowerment enjoyed by the Interstate Commerce Commission under the Woodrow Wilson administration was more subdued under the more conservative administration of Warren Harding. Republican Party ties between William Willcox—the chairman of the Port and Harbor Development Commission and former chairman of the Republican National Convention—and President Harding did nothing to solidify or extend the initial levels of political support the harbor commission had enjoyed. Other local groups potentially affected by the *Comprehensive Plan*'s freight subway proposal—the port region's lightermen, truckers, porters, and manufacturers—either eyed the Port and Harbor Development Commission's plan with suspicion or, despite the plan authors' efforts to publicize the scheme through media releases, lectures, and film showings, ignored the plan altogether.[28]

By 1925 the city's banking and real estate leaders were throwing their weight behind the studies, already under way, of the Regional Plan Association's scheme, which proposed, in sharp opposition to the 1920 Port and Harbor Development Commission's *Comprehensive Plan*, the removal of nearly all existing freight-handling facilities from the center of the city. By the end of the decade, the *Regional Plan* authors were also voicing their opposition to the just-completed West Side Elevated Highway, which had been built with municipal funds (pushed through by Tammany-backed borough president Julius Miller) over the objections of the Wall Street–backed regional planners.[29] The West Side elevated freight rail line, running mostly parallel with the new highway, was built mostly thanks to City Hall (as opposed to Wall Street) support as well—in part as an attempt to compensate for the nonrealization of the Port and Harbor Development Commission's urban freight-handling scheme a decade earlier.[30]

In 1921 the Port and Harbor Development Commission was renamed the Port of New York Authority. Still charged with the task of developing transport infrastructure throughout the region, the Port Authority made several further attempts over the next few decades to introduce new freight-handling facilities and techniques to the port region's urban core, such as a scheme in the late 1940s to build a union car float terminal at 23rd Street, another in the late 1940s to build the largest food terminal in the world, along the West Side waterfront, and a proposal in the early 1960s to build a container-handling complex in west Midtown.[31] As with the 1920 Port and Harbor Development Commission scheme, none of these proposals garnered sufficient political or financial backing. Rather, during this period the Port Authority had far more success developing toll roads, passenger transit lines, airport facilities, and suburban container compounds in New Jersey.[32]

If Fitch's analysis of planners' bias against manufacturing activity in the city begins with rampant speculation in office building during the late 1920s economic bubble, the related history of elite rejection of significant urban freight transport proposals begins noticeably earlier, and so implicates the extent to which such a bias may also have been linked with elite anxieties dating back to at least as early as the 1910s. Thus, a closer examination of such concerns and anxieties during this decade, and the relevance of these anxieties to subsequent trends in urban planning, can begin to answer the question of why, over much of the twentieth century, New York's political and financial leaders were eager to subsidize urban office growth through, among other things, investment in passenger transit facilities, but not to subsidize urban manufacturing growth through investment in urban freight-handling facilities. Such an inquiry may also begin to answer the related questions of why, after World War II, the city's manufacturing sector, despite its established strength in numerous industrial sectors, found itself largely excluded from the lucrative military-industrial contracts that dominated the postwar American economy, and why a planning paradigm favoring the use of urban centers for manufacturing power never emerged. Over the course of the 1910s a series of violent events occurred in New York that, whether directly or through the mediation of significant texts circulating among authorities, helped shape and alter top-down perceptions of freight transport in the city.

The Chenangoes of *Throttled!*

It was John B. Trevor, a longtime Rockefeller friend and influential arch-conservative anti-immigration lobbyist, who insisted that copies of Thomas J. Tunney's *Throttled! The Detection of the German and Anarchist Bomb Plotters* be distributed to the members of the anti-communist Lusk Committee convening in Albany in August 1919.[33] Trevor would go on to found an advocacy group known as the American Coalition of Patriotic, Civil and Fraternal Societies, which by the 1930s was publishing maps of Manhattan claiming to identify the locations of political meetings associated with anti-capitalist newspapers.[34] Trevor's attention in 1919 to Tunney's memoirs likely stemmed from a similar preoccupation with the city's apparent geography of subversive political activity. But if the American Coalition maps portrayed the meeting places of radical newspapers as the "pivots" within such a geography, Tunney's memoirs implicated in this role the city's vessels of clandestine transportation.

The Lusk Committee (so named by reporters for the state senator, Clayton Lusk, who acted as its chair) was officially the New York State Joint Committee to Investigate Seditious Activities. It was backed by numerous New York business leaders, such as J. P. Morgan and John D. Rockefeller and by anti-subversion officials in Washington, such as Attorney General Mitchell Palmer.[35] Such figures sought the application of the legal framework established by the Espionage Act and anti-syndicalism laws of the 1910s—which had in effect outlawed all spoken criticism of the U.S. government, especially of government's role in facilitating wartime profiteering—to an anti-subversive state offensive in New York.[36] The Lusk Committee's backers aimed in particular at the disintegration of the leadership of the state's communist, socialist, and anarchist circles, the raiding of left-wing schools and publishing houses, and the deportation of immigrants suspected of harboring radical, anti-government political ideologies. Indeed, in the same month in which they read Tunney's *Throttled!*, members of the Lusk Committee would order a raid of the Rand School of Social Science in New York, which was the Socialist Party of America's principal institution and organ for workers' education and political dissemination. And five months later, the published reports of the committee would embolden the New York State Assembly to expel five of its socialist members, recently elected from some of the state's most radical districts—New York City's Harlem, the Lower East Side, and the South Bronx—on the grounds that these assemblymen's socialist politics constituted a form of treason (a decision controversial enough at the time to draw sharp criticism from Governor Al Smith, as well as former governor Charles Evan Hughes).[37]

Tunney's book wasn't about schools or political assemblymen, nor did it delve into the legalistic intricacies of linking certain political ideologies to treason. Rather, it recounted his experiences as captain of the New York City bomb squad from 1914 until early 1919 (when Tunney retired, citing physical injury from the job) in "tailing" or "shadowing" the illicit manufacture and transport of bombs and other weapons by subversive groups around the port.[38] If the Rand School raid and State Assembly expulsions reflected a pressing desire, on the part of authorities and financial elites, to eliminate certain "heads" of the working-class radical political movement by undermining several of the movement's nodes of organization and dissemination, the Lusk Committee's attention to Tunney's memoirs reflected something else: a desire to understand and, where necessary, break up a network of subversive traffickers and shadow manufacturers, the specter of which Tunney linked to the very transport geography of the Port of New York.

The picture Tunney painted, of a port city populated by many illegibly mobile, subversive social elements, is rich in its social as well as spatial complexity, and merits a close reading. Tunney portrays the port's bomb traffickers and illicit manufacturers as associated with numerous politically hostile groups active in New York. For instance, there were the Germans. During the United States' wartime neutrality period, lasting from 1914 until 1917, the Port of New York was rife both with German agents seeking to barter for American arms and with German sailors whose ships couldn't depart the port owing to the British naval blockade in the Atlantic.[39] Throughout 1915 and 1916, a flurry of ship explosions and munitions factory arsons occurred throughout the harbor, with German saboteurs often suspected, though rarely proved, to be the agents of destruction.[40]

In Tunney's portrayal, there were also the anarchists, in particular such terrorist groups as the Brescia Circle, named in honor of the Italian American from Paterson, New Jersey, with anarchist leanings who had traveled to Italy in 1900 and successfully assassinated the Italian king. In 1914 and 1915 the Brescia Circle had been involved in the secret movement of numerous bomb-making ingredients around the city—sulfur, sugar, chlorate of potash, antimony, tin, nitroglycerine, TNT, and dynamite—and the plotting of terrorist attacks against cathedrals, police stations, and courthouses throughout the city. Tunney describes the Brescia members as a "cosmopolitan lot"—including not just Italians but also Russian gentiles, Russian Jews, Germans, Austrians, Spaniards, and Americans, all of both sexes, numbering roughly six hundred members total.[41]

Tunney also draws attention to the Indian revolutionaries. Led by the organizer Lala Har Dayal, these anti-imperial insurgents had agents and outposts on at least three continents, their networks stretching from Singapore and Tokyo to Amsterdam and Berlin, as well as to major North American ports and American university campuses.[42] In Tunney's telling, in 1915 Indian revolutionary cells in New York were busy stockpiling "rifles, field guns, swords and cartridges" in a warehouse on Houston Street in

Manhattan for secret shipment to Karachi to ignite an anti-imperialist uprising against the British Raj.[43]

Finally—and most important, in Tunney's account—there were the trafficking links and networks of allegiance among these groups, links born at times of political expediency and at other times of solidarity, and implicating a specter of subversive associations extending far beyond the fluid spaces of the harbor itself. If at one point in his text Tunney writes extensively on the prominent American anarchist leaders Emma Goldman and Alexander Berkman, it is less to suggest that the assassination, imprisonment, or deportation of these two leaders would be a major blow to the specter of anarchism and more to intimate the extent to which the Brescia Circle in New York was entangled and enmeshed within a wider, global conspiracy of radical anarchism. Berkman's radical newspaper *The Blast* interested Tunney not for the principles of organization and specific platform items espoused there but rather for this quotation: "The breath of discontent is heavy upon this wide land. It permeates mill and mine, field and factory. Blind rebellion stalks upon highway and byway."[44]

In this same vein, Tunney mentions a speech delivered by Leon Trotsky at Beethoven Hall in New York City in 1915 less to dwell on Trotsky's importance as an intellectual and organizer for the international Communist movement and more to highlight a phrase from another speaker at the event that intimated the likelihood of German-Bolshevik cooperation.[45] Tunney similarly saw the Indian revolutionaries as centrally implicated in some hazily defined network of collaboration and conspiracy among oppositional groups. For instance, pointing to a wartime letter intercepted by his detectives, Tunney highlights a conspiratorial link running between Indian nationalists in Germany and anarchists in North America, mediated by socialist circles in Holland:

> Dear Comrade,
> Can you send me some earnest and sincere comrades, men and women, who would like to help our Indian revolutionary movement in some way or other? I need the cooperation of very earnest comrades. Perhaps you can find them in New York or at Paterson. They should be real fighters, IWWs or anarchists. Our Indian party will make all necessary arrangements. If some comrades wish to come, they should come to Holland.

The letter, supposedly penned by Har Dayal (though it gives the impression of possibly being a forgery), reads in the postscript, "Kindly be very careful in keeping everything secret and confidential. When comrades arrive they should go and see Domela Nieuwenhuis" in The Hague—Holland's leading socialist politician—"he will tell them where to meet me."[46]

Such was the specter of international intrigue and subversive, clandestine stratagem that Tunney saw as responsible for the illicit manufacture, trafficking, and detonation of bombs around the Port of New York during the course of his tenure as the city's bomb squad captain—a "hydra" of Germans, anarchists, Indian nationalists, Bolsheviks, anti-imperial insurgents, and revolutionaries, connected less by any shared ideological position (indeed, Tunney suggests that for the most part, these groups were hoodwinking each other ideologically[47]) than by a shared agenda of trafficking and detonating arms and explosives and toppling a political status quo. For Tunney and the bomb squad, this agenda was centrally bound up in the very physical spaces and peculiar social divisions of labor in the port itself. Tunney describes how, when a wave of street bombing incidents began to plague New York in 1914, followed by a wave of mysterious ship fires in the harbor in 1915 and 1916, the bomb squad's principal strategy of detection did not involve the shadowing of any German, anti-imperial, or anti-capitalist propagandists, publishers, or intellectuals. Rather, to shed light on the port's networks of bomb makers and bomb setters, Tunney sent his men to the docks, with orders to shadow the harbor dwellers and mobile workers there. In the early twentieth century, these workers numbered in the hundreds of thousands. As late as the 1930s, many of them lived as families aboard moored vessels, which were clustered into "floating villages" or "movable suburbs," especially along the waterfronts of Jersey City, Brooklyn, and the Battery.[48]

In his characteristically evocative language, Tunney explains his detection strategy within this waterside setting, emphasizing the forms of clandestine mobility enabled by the harbor's physical spaces and modes of social organization as the only qualities binding together an otherwise socially dizzying and politically disorienting web of subversive groups:

Anyone familiar with the waterfront of a great port can appreciate its difficulties as an area to be policed.... The contours of the shoreline are irregular, following usually the original margins of solid ground lining the natural harbor, and for every thoroughfare which can pass as a street there are a dozen or two alleys, footpaths, shadowy recesses and blind holes. Locks and keys and night watchmen will protect the land side of the piers, but from the water side entrance to any pier is easy, concealment still easier, and flight no trick at all ... [New York's] waterfront was physically clean and its longshore population, thanks to a competent police force, manageable. And yet, as Shakespeare said, "there are land rats and water rats."[49]

Tunney's aim is to track down these "land rats and water rats" (the reference is to *The Merchant of Venice*[50])—the mobile, casual, and at times criminal elements within New York Harbor's labor force, on the boats and piers

Figure 4.6
A "queer floating village in the City of New York" (caption from original text), 1902.
Source: Frank Leslie's Illustrated Newspaper, June 12, 1902, 567.

of the city's waters, or inside the warehouses, factories, and garages of the city's broken-up masses of land—tail them, and discover how their movements might be opening up possibilities for the circulation and planting of bomb materials throughout the port. After a dramatic flurry of unexplained ship explosions and bombing incidents in 1915, Tunney, noting that most of these ships had been carrying a cargo of sugar, begins to suspect that bombs are being slipped into sugar bags at one of the city's major sugar refineries.[51] Sending his men into the refineries themselves might arouse suspicion and scare the culprits into hiding, so the bomb squad director has his men instead tail what he calls the port's "sugar captains," men who own their own small watercraft, known as "lighters," and, along with a boat crew, make a living moving sugar bags from the big refineries to coastal and oceangoing cargo ships.[52]

Assuming various disguises related to harbor work, Tunney's detectives watch the sugar bags move from refinery workers to lightermen. Working from his agents' reports, Tunney begins to suspect that the illicit cargo transfer is occurring during the night. As he explains, this possibility creates a significant setback for the bomb squad, none of whom are sufficiently acclimatized to harbor navigation to surveil the nocturnal waters and recesses.[53]

Tunney's conviction that the key devices enabling the flurry of bombings around the port are so many small harbor watercraft, darting about the port by night, is reinforced by a break he and his team catch, when the French military attaché telephones the bomb squad about a man known to be going to various factories and wholesalers throughout the city, trying to purchase TNT.[54] This lead sends Tunney and his men on a sojourn through the port region's illicit market for many incendiary materials: not only TNT but also dynamite, potash, coal tar, nitroglycerine, and other dangerous urban factory materials and potential bomb-making ingredients. As the team discovers, the materials often originate from the many explosives factories and "dynamite stores" speckling the western, New Jersey, side of the harbor, yet they can as easily have come from paint and pharmaceutical plants, or metal trade shops, in Manhattan.[55]

Tunney has several of his agents disguise themselves as the operators of a dynamite store in Perth Amboy. From this vantage point the detectives discover a "Do-Do Chemical Company" in Manhattan—a front for chemists and chemical smugglers sympathetic with the German cause[56]—a Manhattan clockmaker with unusual delivery orders for nitroglycerine, and, finally, a man going by the name of Robert Fay and posing as an inventor in a garage in Weehawken, New Jersey, just across the Hudson River from

Midtown Manhattan.[57] Tunney's undercover detectives find that, through an elaborate chain of hand-offs, the TNT and clocks make their way to Fay in Weehawken, who secretly assembles timer bombs in his garage and stores them in a wooded shack on the outskirts of town.[58] Tunney and his men find among Fay's possessions a "chart of New York Harbor" and information that Fay has a motorboat at a slip on the Hudson River, opposite 42nd Street. Tunney expresses his feelings of vindication on discovering the chart and boat: this "pointed the finger of guilt toward the waterfront which after all those months of waiting was the direction in which we were most interested."[59]

Emboldened now by a renewed sense of the secret movement of materials about the harbor, Tunney turns his attention back to the local class of watercraft pilots. Intrigued by Fay's use of a harbor slip to move undetected about the port, and still maintaining strong suspicions about the lighter captains carrying sugar, Tunney fixates on another denomination in the harbor's peculiar social geography. He recalls in his memoirs that during his investigation of the sugar lightermen, he and his detectives "made the acquaintance" of a category of harbor workers known locally as the "Chenangoes." These were, writes Tunney,

nothing more romantic than flyby-night stevedores whom the lighter companies engaged at the sugar wharves to load cargoes. They worked by the day, or by the job, there were always plenty loitering around to be hired, and they drew their pay and went their way. No one ever had to wonder who they were or where they came from, for a stout body was all the recommendation a Chenango required.[60]

What intrigues Tunney about the Chenangoes is their day-to-day mobility, their propensity to "disappear" from the city for stretches of time, migrating along the coast looking for work. Tunney is also fetched by the Chenangoes' national and linguistic diversity. They are, mused Tunney,

[a] type of common labor, the same, I suspect, that carried materials for the Tower of Babel, and speaking almost as many tongues. The same face rarely appeared a second time to be hired—not that there was anything particularly unpleasant about the work, but rather that all work is repulsive to a Chenango. He is the hobo of labor and if the same man had been re-hired, no one would have noticed or cared. We paid such attention to them as their variety permitted—followed them to all the points of the compass.[61]

A colleague of Tunney's, one familiar with the sabotage methods "taught in the Wilhelmstrasse," advises him that the odds that German spies are indeed making much use of the New York Harbor Chenangoes are slim, and anyway "the arrest of a guilty Chenango would not supply the background necessary to picture the bomb system in its entirety," for "the destroying agent rarely knows the men higher, the real conspirators."[62] But of course, Tunney has many other centers of conspiratorial planning to worry about than just the Wilhelmstrasse—there are the Indian nationalist leaders in Amsterdam, Tokyo, and Berlin, the anarchist and Wobbly cells in New York, Chicago, Toronto, and London, the Bolsheviks in Moscow and Saint Petersburg, and so on. Dismissing his friend's advice, Tunney has the bomb squad return to the sugar theory and the "floating tribe," as Tunney also calls the Chenangoes, "who came, worked once, and vanished," and "were of all nationalities."

Tunney spends much of 1915 having his men follow this group, asking "enough questions about them to learn the entire history of any civilized people."[63] And his team does indeed uncover one bomb conspiracy, against the steamship *Kirkoswald*, involving a Chenango. Tunney describes how a German chemist in New York by the name of Scheele designed several bombs in his laboratory in Manhattan and met with German agents at the Brooklyn Labor Lyceum to hire "small fry," or Chenangoes, to slip these bombs into the sugar bags bound for the *Kirkoswald*'s hold. Scheele had delivered several of the "infernal machines" from his lab personally, smuggling them in a bag of sawdust under his arm. Another agent took the bombs piecemeal to waterfront saloons along the West Side, and from there to destroying agents at the White Star piers. The conspirators found a Chenango to do the hull smuggling work in May, but the *Kirkoswald* bomb was found by sailors before it could be detonated. One of the German agents apprehended by Tunney claims that this same bomb-setting technique—entailing the movement of bombs from illicit lab to Labor Lyceum to waterfront saloon to Chenango to ship hold—had been used to smuggle bombs aboard the American ship *Lusitania*, but that, coincidentally, the German navy had torpedoed the ship (thus instigating the American entry into the war) before the "cigars" could ignite.[64]

In his memoirs, Tunney's preoccupation with the local social category "Chenango" hinges on such investigations. It might also have reflected a degree of local knowledge on his part regarding the term's history and connotations in the port. In published writing, Tunney's is among the first uses of the term, though there are references to New York Harbor's "shenangoes" in Charles Barnes's sociological studies of New York City longshoremen between 1910 and 1915.[65] In dictionaries that carry the word (and there are only a few), the origin of the term "Chenango" is obscure. Two relatively recent dictionaries hazard a guess that the term refers in some way to the

Chenango Canal, a branch of the Erie Canal that connected Utica, New York, to the Susquehanna River between 1834 and 1878. Neither dictionary author presents any evidence supporting this theory.[66] Nonetheless, to the extent that this etymology highlights the history of social and political associations between New York Harbor dock labor and upstate New York canal labor, it would be an error to reject the theory outright. The Chenango Canal, like many of the other canals built in the American Northeast during the nineteenth century, was constructed and later operated primarily by Irish American navvies. These mobile labor armies of diggers and pullers were notorious for their drinking, their rough, violent culture, and their involvement in sectarian gangs such as the Orangemen and their rivals, the Whiteboys and the Ribbonmen.[67] Feuds among these gangs, sparked as much by local conflicts over work and resources as by religious differences and opposing ideas about the future of Ireland in the British Empire, often erupted in riotous violence at the major nexus points in the Northeast's waterway network. Such conflicts included the bloody Orange Riots, which broke out between Orangemen and Ribbonmen in Manhattan in 1870 and again in 1871.[68]

The Chenango Canal closed later in the 1870s, the first such closure in the history of the New York State canal system. It is at least plausible that many of the Chenango's barge families migrated to New York Harbor looking for work, that the unprecedented appearance of canallers from a closed canal line made a strong local impression around the harbor, and that soon thereafter the term "Chenango" came to refer, in local jargon, to any casually employed, semi-migratory harbor worker. None of this seems to turn up in any archival record, though—not even a migration of Chenango Canal boatpeople to New York City after the closure of the line in the 1870s. Such an etymology is perhaps too tenuous on its own.

An alternative, or complementary, etymology (and it is plausible that the two are not mutually exclusive) and one that, much like Tunney's memoirs, associates harbor workers with harbor bombings, could instead begin with a mysterious ship explosion in New York Harbor in 1864. During the American Civil War, numerous steam-powered gunboats were built at the Brooklyn Navy Yard, on the Wallabout Bay arm of New York Harbor. Many of these boats, such as the USS *Oneida*, the USS *Genesee*, the USS *Onondaga*, and so on, were named after counties in New York State and subsequently sent southward to blockade and bombard the bays and river mouths of the Virginia and Carolina coastline.[69] Among this fleet was the paddle-wheel gunboat *Chenango*. It was bound for Fort Monroe, Virginia, but never made it out of the harbor. On April 15, 1864, the *Chenango*'s pot boiler exploded with "terrible violence," killing thirty-two soldiers and injuring dozens of others. Officials called the blast an accident and dismissed the ship engineer, who survived the incident, for negligence. Locally, suspicions that the explosion had been the result of a Confederate spy armed with an "infernal machine" of some kind were widespread.[70] If the *Chenango* explosion was indeed the work of a saboteur, it was the only ship belligerently attacked in New York Harbor between the War of 1812 and World War I.

Earlier in 1864, Union Army officials had intercepted a letter from a Confederate spy named Alexander Courtenay boasting to a friend about an invention of his, soon to be adopted by the Confederate Secret Service, called a "coal torpedo"—a piece of cast iron molded to resemble a large clump of coal, hollowed out and filled with gunpowder. "The President [Jefferson Davis] thinks them perfect," effused Courtenay.[71] Wary of the draft riots and specter of violent sedition already under way in New York City, Union officials did not expose the intercepted letter to the public. Nonetheless, New Yorkers at this time were likely already very aware that the city was under some kind of subversive assault, though the question of whether the assailants were Confederate agents, draft rioters, or a conspiracy between both likely appeared hazy. Waterfront arsons were common enough in 1863 and 1864. Several decades later, one former Confederate spy described arriving in New York, receiving bottles of "Greek fire," a burning liquid, from a local chemist (also a spy), and taking the bottles to a Hudson River wharf to set "vessels and barges of every description" on fire.[72]

In December 1875, public suspicions that the Port of New York had indeed been attacked by Confederate agents were confirmed in a letter to the *Times of London*, which was reproduced in the *New York Daily Herald* a few weeks later.[73] The author of the letter identified himself only as "E" and claimed to have written the letter in Bayswater, England—pointing to the likelihood that the writer himself was a former Confederate spy who had fled to Britain after the war. E wrote to *The Times* in response to that paper's coverage of a large explosion on the waterfront of Bremerhaven, Germany, earlier in 1875, which had killed some eighty people—the work, German investigators concluded, of a transatlantic confidence man who took out insurance policies on ships and then slipped "coal torpedoes" into their cargo holds.[74] Roused by this news item, E wrote *The Times*:

Perhaps a short account by one who is acquainted with the origin and object of the coal torpedo, referred to in the telegram from your Prussian Correspondent in *The Times* of yesterday, may interest some of your readers. In the winter of 1863 a Captain Courtenay obtained from the Confederate War Department the facilities for making some experiments with his coal-castings at Richmond, Virginia, to be used for collapsing the boilers of the enemy's war vessels and transports by introducing the sham coal into the Government coal piles at the various United States Naval depots along the coast, and on the great rivers of the interior.[75]

The writer revealed a good deal of intimate knowledge about the construction and design of these "sham coals": "the shell being filled and closed with a brass plug, it was dipped by means of a string into a boiling mixture of coal tar, pulverized coal." According to E, the spy Courtenay and his Confederate Secret Service agents had smuggled some of the torpedoes into a U.S. government coal depot. As E highlights, this "led to the destruction of the new sloop of war *Chenango*," as well as of the "splendid steamer *Greyhound*," off the coast of Virginia.[76]

E ended his letter by stressing that the manufacture and distribution of coal torpedoes did not end with the American Civil War. To the contrary, E pointed out that an entire industry and market throughout the North Atlantic had emerged around the device—that "now, in 1875, the United States Navy is ahead of all others in the science and practice of offensive torpedo warfare," and that, moreover, it was

proposed by a Committee of the Fenian Congress assembled in New York in 1867 to distribute a quantity of coal torpedoes in the coal cellars of her Majesty's Ministers and prominent public men in London, but was abandoned on the suggestion that many kitchens in London would have Irish maids connected with the scullery department.[77]

What E's passing reference to the momentary Fenian flirtation with the coal torpedo suggested was that Confederate spies were not the only agents capable of wielding and employing this destructive "sham" technology; nor, for that matter, was the civil war that had just occurred in America the only form of violent civil conflict in which such technologies could come into play. Indeed, in the United States during the 1870s, civil fears that had once been fixed on the specter of Confederate secession instead became fixed on the specter of an armed working-class revolt. One 1871 editorial in the American journal *National Quarterly Review* recommended that decades-old military engineering plans to wall and fortify American cities be followed through, not so much because slavery might once more "become a bone of contention" as because "free trade and protection may come into collision.... We can even imagine a Communist war, against property, and general uprising of Labor against Capital, as within the bounds of possibility."[78]

The form of urban fortification this editorialist had in mind was relatively conventional—for instance, a chain of redoubts, originally proposed in 1858, to divide Brooklyn and Queens from the rest of Long Island (lest some enemy fleet land at Sag Harbor, or an insurgent force emerge among the farmers and fishermen of Nassau County).[79] Instead, elites in New York City, disturbed by revolutionary events in Paris several years before and by the tumultuous first general strike of American railroad workers in 1877, invested in a large, privately financed armory in the city's wealthiest district, the Upper East Side, to defend residents of the area against a potential uprising by the city's "dangerous classes." J. P. Morgan Sr. contributed money to the project, a "sure guarantee for the future," as he wrote.[80]

Outbursts of violent class warfare had occurred in New York City before, whether in the form of the draft riots of 1863 or the Negro Conspiracy of 1741. The arrival of Johann Most into the city in 1882 introduced to this history, as if providing a foil for the freshly completed Upper East Side Armory, a new generation of violent and subversive weaponry for working-class insurgency. An anarchist organizer deported from Germany, Most advocated "propaganda by deed," and sought to transform the substance of dynamite into a violent political symbol.[81] Most picked up knowledge of bomb-making and incendiary materials while working at an explosives factory in Jersey City in 1883. Entering into anarchist circles in New York City, Most gained two loyal young followers in Emma Goldman and Alexander Berkman, who helped him publish and distribute his German-language pamphlet *Revolutionäre Kriegswissenschaft* (Science of Revolutionary Warfare), which advocated the theft and guerrilla manufacture of dynamite, nitroglycerine, gun-cotton, fulminating mercury, and poisons, using chemical by-products such as coal tar from common urban industries.[82] After the Haymarket bombing in Chicago in 1886, many officials suspected that the assailants used this pamphlet as a manual for action.[83]

The American union leader Samuel Gompers would later date the split between conservative unionism, associated with Gompers's own American Federation of Labor, and revolutionary unionism, associated not only with violent radicals like Most but also with the "one big union" agenda of the Industrial Workers of the World, to the 1886 bombing in Chicago.[84] In Gompers's historical assessment, once this schism was opened, class relations in the United States became increasingly belligerent, marked on the one hand by massacres of striking workers' picket lines and on the other by numerous assassinations and attempted assassinations of financial and political leaders.[85]

This pattern of violence directed against the city's wealthy class and business leaders continued through the 1910s. In 1914, only six days after the assassination of Franz Ferdinand on the banks of the Miljacka, a townhouse on Lexington Avenue in New York exploded, a botched anarchist plot to assassinate John Rockefeller.[86] The following summer, an assassin shot and wounded J. P. Morgan, though this was not an anarchist but rather a German sympathizer trying to intimidate the powerful banker into discontinuing his financing of and profiting from the war in Europe.[87]

After the Lexington Avenue explosion, the New York police commissioner announced the creation of a first-of-its-kind, undercover urban bomb squad to infiltrate the city's underground or seditious political circles, and placed Tunney in charge of the outfit. The flurry of anarchist bombings continued through the remainder of 1914: St. Patrick's Cathedral and two separate attacks on St. Alphonsus Church in October, the Bronx County Courthouse and the Tombs Police Court in November.[88]

In early July 1916, a bomb went off at the front door of the New York City police headquarters on Centre Street, an incident Tunney never mentions in *Throttled!* but was surely weighing on his mind.[89] On July 30, a massive explosion shook the long wharf at Black Tom Island in New York Harbor, at that time the largest munitions depot in the United States. This event, which killed seven munitions wharf workers, is not mentioned in *Throttled!* either—officials immediately called it an accident and did not send the bomb squad to investigate the incident—but shortly after the publication of his memoirs Tunney was asked by the Bureau of Investigation to help interrogate one key German suspect captured in El Paso.[90] Eyewitnesses of the Black Tom explosion said they had seen a fire start on a harbor barge called *Johnson No. 17* and spread from there to the munitions and explosives along the wharf until culminating in a furious blast, powerful enough to shatter office windows in midtown Manhattan, four miles away. The culprits were never caught, but police authorities suspected the barge had been set alight by anarchists or German saboteurs.[91]

To return, then, to Tunney's choice, writing his memoirs in 1919, of the word "Chenango" to describe the class of mobile harbor people most preoccupying his men during his tenure as captain of the New York City bomb squad: we should not overlook the possibility that Tunney, directly or indirectly, was connoting this entire history, stretching in time from the USS *Chenango* disaster of 1864 to the Black Tom disaster of 1916, of clandestine bombs, bomb materials, and bomb makers moving about the port. Just so, for some members of the Lusk Committee reading the freshly published *Throttled!* at the lobbyist John B. Trevor's behest, the association (however Tunney may have intended it), between New York's "Babel" of mobile harbor workers and a deeper political history of violent civil and class warfare, both within the dark recesses of the port and the belligerent geographies beyond, was likely hard to miss.

On September 16, 1920, nine months after the New York State Legislature, perturbed by the Lusk Committee reports, had expelled five socialist members from the New York State Assembly, a bomb went off outside the Stock Exchange at the corner of Wall Street and Broad in Lower Manhattan. Bomb netting already covered most of the windows of the banks and financial buildings of the district. J. P. Morgan, whose bank was opposite the Stock Exchange, was likely the intended target of the attack. He wasn't at the office that day, but his son, Junius, was. Junius was injured in the blast, along with scores of accountants, receptionists, and delivery boys.[92] When the smoke subsided, more than three dozen people at Wall and Broad were dead. As with the USS *Chenango* explosion, the harbor boat explosions of 1915, and the Black Tom blast, the true cause and culprit of the Wall Street bombing were never discovered, though anarchists seemed the likeliest suspects. Pink fliers had been found earlier that day at a Hudson River terminal. "Remember," the fliers read, "we will not tolerate any longer. Free the political prisoners, or it will be sure death for all of you.—American Anarchist Fighters." Yet no link between these fliers and the bombing at Wall Street was ever proved.[93]

A hunt began, spearheaded by Morgan, Rockefeller, Attorney General Palmer, Bureau of Investigation director William Flynn, and the young intelligence director at the Justice Department, J. Edgar Hoover. These officials' primary suspect was a group of Italian American anarchists in eastern Massachusetts known as the Galleanisti, named for their leader, Luigi Galleani, who, much like Johann Most a generation before him, had arrived in the United States advocating propaganda by deed. In Massachusetts Galleani had published the radical circular *Cronaca Sovversiva*, as well as a small, innocuous-looking pamphlet titled *La salute e in voi!* (Health is in you!), which appeared on the outside to be a medical self-help booklet but was in fact an instruction manual, written entirely in Italian, for making nitroglycerine bombs using materials pilfered from chemical factories.[94] Galleani had been deported in 1919, and his followers were known either to have followed him to Italy or to have gone into hiding in various East Coast ports. To track the Galleanisti, Bureau of Investigation director Flynn turned to New York City's waterfront and longshoremen, just as Tunney had done a few years before to track down Brescia Circle and kaiserite bombs. The Bureau of Investigation established surveillance programs throughout the East Coast ports, hoping to find information that would lead them to the Wall Street bombers. They found stowaways, but no Galleanisti.[95]

It was also in 1920 that the New York, New Jersey Port and Harbor Development Commission published its *Comprehensive Plan*, advocating infrastructural investment in the Port of New York—in particular the borough of Manhattan, including the Wall Street area—as a place where physical materials could be delivered, assembled, circulated, and dispatched. As the fate of the *Comprehensive Plan* over the course of the 1920s suggests, financial and political elites had other ideas. After World War I, some leaders had shown enthusiasm for enacting the recommendations of the *Comprehensive Plan*. But after 1920, a year that saw the last of a long succession of New York City bombings (at least for a time) and also the first such bombing aimed at people on Wall Street, that enthusiasm rapidly unraveled, and elite support gravitated to another, quite different plan for the future of freight in the city.

Casual Harbor Work, Shadow Manufacturing, and Comprehensive Planning

The Regional Plan of New York and Its Environs, one of the largest comprehensive planning studies in American history, was released to the public in 1929. The result of economic research and planning studies dating back to 1922, the report comprised of more than a dozen volumes, as well as a shelf's worth of graphical surveys and supplementary studies, dealing with questions of zoning and land use, as well as planning issues ranging from highway and transit layout to airport facilities, civic architecture, and housing development in the urban core. Drawing on two generations of architects and planners, the *Regional Plan*'s visual proposals combined the neoclassical Beaux Arts sensibilities of fin-de-siècle American planning with the more modern, streamlined Art Deco forms that would come to eclipse the Beaux Arts style in popularity by the 1930s.[96] The plan was financed by the Russell Sage Foundation, named for the prominent New York City industrialist and banker who had died in 1906 (Sage had survived an assassination attempt in 1891). On the board of the Regional Plan Association sat representatives from Morgan Bank and First National. The oil magnate and local real estate baron John D. Rockefeller would also exert his influence on the recommendations of the plan.[97]

The *Regional Plan* was not the first in-depth study of New York City the Russell Sage Foundation had funded. A decade before the first *Regional Plan* studies, the foundation—much like Thomas J. Tunney—had been interested in casual, "floating" dock and harbor labor around the Port of New York. Between 1910 and 1915, the foundation funded Charles Barnes's sociological studies of the organization of longshore work around New York Harbor, publishing and distributing his final study, *The Longshoremen,* in 1915.

In this report, Barnes was at pains to distinguish between the "regular" longshoremen of the port, who worked regular hours and enjoyed consistent, company-based employment on the docks, and "irregulars," who worked only at odd intervals throughout the year, showing up when they could to a wharf, amassing in "shapes" outside a pier warehouse, hoping to be picked by a stevedore foreman for a day's wages.[98] Barnes refers to this latter group as the harbor region's "shenangoes." The descriptive contrast between Barnes's shenangoes and Tunney's Chenangoes is striking. If Tunney's Chenangoes were a "Babel" of nationalities, Barnes' shenangoes were

largely recruited from Irish longshoremen, either broken down by drink and unfitted for regular work, or longshoremen in good standing temporarily out on a spree. The Irishman, who stands as the best type of waterfront worker, when he begins to lose ground slips to the lowest level and represents the worst.[99]

Similarly, if Tunney, writing in 1919, asserted that "a stout body was all the recommendation a Chenango required," Barnes in 1915 cast the shenango as no more than a dockworker turned drunkard, who, finding his "muscles growing soft … little by little he becomes unfitted for the work, is no longer taken on, and finally falls into the class of the shenango." Barnes blamed this group of "drunken shenangoes loafing about the pier entrances" for widespread negative public attitudes toward and perceptions of "waterfront people" in general. Advocating the shenango group's removal from the division of labor around the harbor, Barnes writes that the "down-and-out" shenangoes may "wear longshoremen's hooks, hang about the waterfront, and handle the cargoes of barges and lighters," but they "must be distinguished from the skilled and semi-skilled workmen whose good name they have tarnished."[100]

In his memoirs, Tunney at times romanticizes the Chenangoes, casting them as the harbor's mysterious corps of "fly-by-night stevedores." Barnes reserves his romanticism for the regulars, associating them with a "spirit of independence" dating back to the burning of the British armed vessel *Gaspée* by Providence longshoremen in 1772.[101] Further characterizing this class of worker, Barnes observes that the regular dockworker is "very shy." Embarrassed by the behavior of his shenango comrades, the regular would hide his own "gnarled hands."[102]

Barnes names other semi-casual groups, besides the shenangoes, who occupied, in Barnes's view, a social position well beneath such "permanent" workers. For one, there were the "banana fiends," specializing in the tortuous handling of cargo from the banana boats in seasonally from Central America. These banana handlers were to be found at Coenties Slip on the

East River, and, "as among the shenangoes, there are many Irish and Irish Americans. Educated, even professional men, it is said, have been reduced to this labor through misfortune and drink."[103] And there was the class of lumber men, "sailors out of work" who would take up lumber handling temporarily at lumber yards in the South Bronx, the West 60th Street car float terminals, the Gowanus Canal, and Newtowne Creek.[104] Lighters, Barnes notes, were often manned both by regular crew and by additional shenango crew. Lighter captains were responsible for the solicitation and hiring of day laborers; the captains visited the "shapes" of men lingering outside the piers and convinced some that they would be better off earning a sure, if relatively low, day's wages on a lighter than waiting around the piers for work that, though better paying, they might not get at all.[105]

Despite these divergent characterizations of the shenangoes, or Chenangoes, by Tunney and Barnes, there seems to be little question that they were referring, more or less, to the same body of workers. Why, then, the different spelling of the term? If indeed Tunney ever encountered Barnes's work in 1915 or earlier, it seems that by 1919, Tunney had chosen to reject the Barnes spelling and assert his own. It is plausible that Barnes, who picked up the word during his interviews of dock supervisors in 1910, misheard it, and that Tunney, familiar with a body of local lore eluding Barnes, and still preoccupied with the potential association between floating harbor workers and floating harbor bombs, saw fit to correct the error in his published memoirs. Both spellings entered into New York State labor relations terminology for several decades thereafter, Barnes's spelling appearing, for instance, in Waldo Browne's *What's What in the Labor Movement* in 1921 as a synonym for "dock-walloper,"[106] and periodically in newspaper articles on waterfront labor strife during the 1940s and 1950s;[107] while Tunney's appeared in several state waterfront commissions investigating the implementation of longshore decasualization, advocated by Barnes as early as 1915 and eventually achieved in the 1960s. After that the term, in either spelling, largely disappeared from published use.[108]

By the 1920s the directors of the Russell Sage Foundation had shifted their focus away from Barnes's sociological research on dockworkers and toward the formation of the *Regional Plan*. While the plan volumes never mention the shenangoes of Barnes's study, they do discuss the waterfront at some length, not so much to make recommendations for the organization of work there as to advocate the sweeping deindustrialization and removal of freight-handling activities from the inner-city waterfront altogether. In place of such activities, the *Regional Plan* advocated new office towers, parks, luxury apartment complexes, and ocean liner terminals.[109]

The *Regional Plan*'s stated justification for such deindustrialization schemes was laid out in the plan's economic survey, directed and authored by the Columbia economist Robert Haig. Wrote Haig on the "economic" rationale behind the plan's agenda of urban deindustrialization:

It was hoped that, by observing what is actually happening in the competitive struggle for urban sites, and existing tendencies in the location of economic activities, it
might be possible to glimpse the outlines of an economically ideal pattern or plan;
that, by examining what is being crowded out of the choice central locations and
what doing the crowding, it might be possible to infer where "things belonged."[110]

Haig's chief observation and prediction along these lines was that "on the whole manufacturing is not more than holding in the center of New York, and has already begun to be crowded outward."[111] Haig explained this "trend" by asserting that certain "functions," such as "managing and administering, buying and selling, financing and risk-bearing, investigat-ing and advising," are by their very nature more appropriate for sites in the center of the city than are other functions, in particular the preparing and assembling of materials, which activities "can be and are carried on more economically outside the urban centers."[112] Conflations like these—"can be and are"—that blend the real, the predicted, and the desirable permeate Haig's pronouncements on urban economy. For instance, when discussing the men's clothing industry in Manhattan—which in the 1920s employed roughly 60,000 people and showed no statistical signs of decline[113]—Haig begins by stating that "some industries like the men's clothing industry are destined to leave but will not and should not leave immediately, as too much investment is wrapped up there." But in the long run, in Haig's assessment, such activities as men's garment making had no place in what New York was "physically to become."[114] They were simply the "temporary occupants of obsolete buildings."[115]

For Haig, the future of the urban center belonged to newspaper busi-nesses, where the element of time was paramount; to the seasonal "style" industries, such as fashion and theater; and above all else to the finance sector, which, in Haig's telling, was "in effect, one big structure" whose "streets, practically cleared of all except pedestrian traffic, are no more than corridors and air-shafts" providing a stage for the transfer of valuable and exclusive financial information among business elites.[116] Freight-intensive manufacturing activities, by contrast—the assembly of wood products, metal products, tobacco and food products, and chemicals—were, in Haig's view, doomed to be outcompeted by the more "appropriate" urban trades for inner-city sites. Haig launched several "proofs" that the urban chemical trades in particular were on their way out, presenting, for instance, a map captioned "Analysis of the Migration of 32 Large Chemical Plants During the Last 25 Years: 1900–1925."[117] Here, many arrows leading from dots in Manhattan to dots in New Jersey appear to illustrate a naturally occurring trend of urban deindustrialization. The map is brazenly misleading. During the period to which the map refers, the actual number of chemical plant workers in Manhattan *increased* by 40 percent and the number of chemical plants by 10 percent.[118] What the map actually shows are only those plants deemed "large" (a term that the map's key does not define) and only those plants that changed location; plants that stayed put during this time frame are not accounted for.

The chemical trade was the only industrial sector for which Haig included such a map. Yet a very similar map might have been included for the financial sector, for which Haig reserved such praise as a suitable inner-city economic activity. As Haig himself acknowledged, between 1900 and 1920, a great many activities within the Wall Street financial sector were forced to "detach" themselves from the district and relocate elsewhere. As Haig hastened to add, "this result has been, not a positive decline of the absolute importance of Wall Street, but a narrowing of the scope of activities and a specialization of the functions performed in this district."[119] In other words, while Haig was eager to point to the gradual "sorting out" within the chemical sector, between plants appropriate for urban sites and plants appropriate for elsewhere, as evidence that *all* the chemical trades were on their way out of the city, he was wholly unwilling to draw a parallel conclusion from the very same sorting-out process occurring within the finance sector. In Haig's assessment, the finance industry's "one big structure" was in the city to stay.

One possible explanation for such analytical bias against urban manufacturing activities, and against urban chemical plants in particular, emerges from Haig's discussion of different groupings within the chemical industry and of the difficulty, from the top-down perspective of an economist or regional planner, of keeping track of the flows of materials between these groupings. Haig distinguishes between "heavy chemical" plants, dealing in acids, potash, ammonia, and sodas, manufacturing explosives, fertilizers, and guns, and clustered along the waterways of the New Jersey Meadowlands, and "fine chemical" plants, many of them located in Manhattan, manufacturing alcohol, essential oils, pharmaceuticals, drugs, medicines, and photographic chemicals. A third group, plants for assembling paints and dyes, appears with equal weight in both the city center and the metropolitan outskirts. Haig highlights the categorization problems posed by

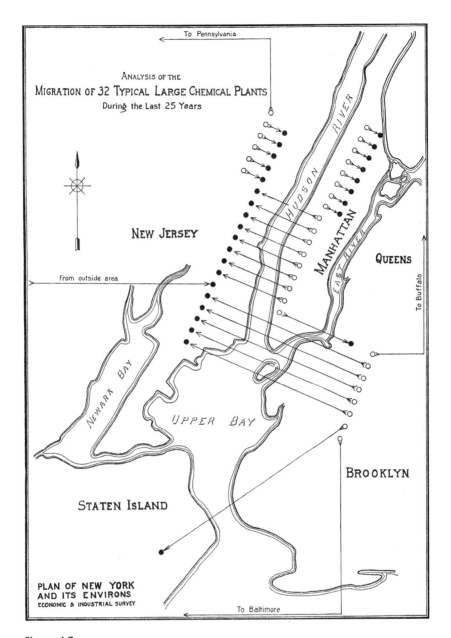

Figure 4.7

A misleading map included in one of the economic volumes of the *Regional Plan.* At face value, the map appears to show an exodus of chemical plants away from Manhattan between 1900 and 1925. In fact, the absolute number of chemical manufacturing workers in Manhattan over this period increased by 40 percent and the absolute number of plants increased by 10 percent.

Source: Robert Murray Haig, *Regional Survey of New York and Its Environs,* vol. 1A, *Chemical, Metal, Wood, Tobacco and Printing Industries* (New York: Russell Sage Foundation, 1928), 8.

certain industrial materials, such as coal tar, that were used for pharmaceuticals and explosives alike, and also the categorization problems posed by many "fine chemical" plants on the West Side of Manhattan and in inner Brooklyn, which sold the chemical by-products of their assembly runs to their "heavier" counterparts in the less thickly settled parts of the port, as well as to many lighter, smaller shops in the center of the city.[120]

In other words, the problem with the urban chemical sector, in Haig's evaluation, may have had less to do with the sector's supposedly failing to "hold" in the center of New York and more to do with the fact that the sector's geographic inner workings, the flow of its materials from plant to plant and pier to pier, were illegible from a top-down perspective. This illegibility is, of course, entirely comparable to the illegibility of the inner workings of the Wall Street financial district's "one big structure"—except, in the case of the chemical trades, the obscured movements were not the flows of financial secrets among banking elites but rather the flows of coal tar, acids, and potash from plant to plant and from sector to sector within the city, from one side of the harbor to the next, carried by porters, Chenangoes, and lightermen. Moreover, they were the flows of the very same materials highlighted in the Galleanisti bomb-making pamphlet *La salute e in voi!*, in Johann Most's *Revolutionäre Kriegswissenschaft*, and in E's coal torpedo letter published by papers in London and New York.

Transshipment of Uranium

Over the course of the 1930s and 1940s, the *Regional Plan*'s 1929 "prediction" of (or insistence on) increasing urban industrial irrelevance and decline was largely negated by actual subsequent trends and events in the history of industrial production in New York. The metal and chemical sectors of the city spawned what was, during these two decades, a relatively vibrant electronics sector, clustered near the docks on Manhattan's Lower West Side, focused on the innovation and small-scale manufacturing of electron tubes, capacitors, radio parts, filters, dials, sockets, antennae, and so on. As a later economic study would contend, New York's mid-century electronics firms benefited from their position as, in effect, "one big structure" of experimental workshops, wholesalers, and suppliers nestled within the city's more broad-based chemical and metal trades, "shopping" each other's front stores for new parts and ideas and drawing needed capital installments from large appliance manufacturers and military contractors.[121]

Haig's insistence on the increasing irrelevance of New York as a center for freight-handling and production was further contradicted by the role the port played in the early stages of the atomic weapons research program launched by the governments of the United States, Britain, and Canada during World War II and culminating in the development and wartime use of the atomic bomb. In 1940 and 1941, the location and shipment of adequate supplies of uranium were crucial hurdles in establishing an efficient atomic research program that would be capable of outmaneuvering German scientists in the race to develop nuclear weaponry. Allied military planners located an initial supply of uranium in Port Radium, northern Canada, near the Eldorado Mining Company's outpost on Great Bear Lake.[122] The movement of this material southward (along the same interior corridor highlighted by Watson Griffin in 1890 as a potential North American inland shipping route to East Asia[123]) required a considerable amount of logistical improvisation. Dog sled teams, driven by Dené mushers, moved sacks of the pitchblende ore from the seams to Port Radium. Next, wooden barges carried the material down the Great Bear River to Fort Norman, then up the Mackenzie River to Waterways, Alberta. The material then went by rail to the Eldorado refinery plant at Port Hope, on the north shore of Lake Ontario.[124]

When the U.S. general Leslie Groves assumed control of the atomic weapons development project in 1942, codenaming the effort the Manhattan Project, he was eager to establish a more convenient uranium supply line, and one less vulnerable to potential enemy attack. He and his officials were amazed to discover that lying undisturbed in a warehouse along the Kill van Kull waterway in New York Harbor were 1,200 tons of Congolese uranium, secretly transferred there by the Belgian mining conglomerate Union Minière at the outbreak of the war.[125] The Belgian executives of Union Minière were immediately contacted by Groves's team and were promised lucrative military contracts for the material. The ore was transported by secret vessels to a pier on the West Side of Manhattan and held at nearby warehouses on West 20th Street.[126] At this clandestine Manhattan location, the ore could be inspected directly by nuclear scientists from Columbia University, then shipped to the Port Hope refinery in Canada. The precise transport logistics of this phase of the operation are not clearly accounted for in available records, but in light of the project planners' decision to use pierside storage facilities, it seems likely the preferred shipping method for moving the ore between New York and Port Hope was by barge: via the West Side piers (the Department of Energy records say Pier 38[127]), the New York State Barge Canal, the canal-lake port of Oswego, and then across the lake to the Canadian shore. From Port Hope, the refined ore was sent in three directions: back to the West 20th Street warehouses, for the benefit of the New York scientists; southward, by rail or barge, to the Seneca

Army Depot in the inland Finger Lakes region; or southward by rail to the "atomic cities" under way, by that point, in Tennessee, Washington state, and New Mexico.[128]

In August 1945, American pilots dropped the "Little Boy" bomb over Hiroshima, Japan; it likely contained uranium sent from both the Great Bear Lake region of northern Canada and from New York Harbor—the two ends of the original "highway of the atom."[129] Shortly after the end of the war, this highway became an immediate source of concern for British and American geopolitical strategists and intelligence officials committed to the post-war military containment of the Soviet Union. As early as 1946, a case of financial fraud and missing ore at the Great Bear Lake mines had erupted into a "black market uranium" scare. The conservative American magazine *National Republic* went so far as to cast the fraud as a treasonous plot to "bootleg uranium to Communist Russia," a charge that would resurface during House Un-American Activities Committee hearings in 1949 and 1950.[130]

This scare concerned uranium from the Far North, yet the problem of surveilling both the movement and discovery of dangerous materials like uranium extended southward as well, to the flow of goods about New York Harbor. Of course, uranium had never been mined from New York, but it had been unexpectedly and fortuitously found there. This was in many ways the very advantage of the Port of New York as an industrial and organizational space: it was a place where all sorts of cargoes, including dangerous, combustible materials, could be found, set aside, and moved about in secret. For Groves, the cargo was uranium; for the anarchists, it was dynamite from the region's agglomeration of explosives factories; for the German saboteurs during World War I, the cargoes included TNT and clock parts from the city's "shadow" manufacturing sector of "Do-Do Chemical Companies" and chemist-imposters.

In a different political context, such usefulness as an organization space, to say nothing of the city's nascent electronics sector and similar innovative industries, might have signaled a potentially critical role to be played, by the port and its configuration of urban industrial spaces, within the military-industrial political economy that was to emerge after the war. It is even plausible that in a less fearful political climate, the urban industrial infrastructure schemes put forward by a previous generation of civil engineers—schemes like the Port and Harbor Development Commission's plan of 1920—might have been rediscovered, dusted off, and evaluated in a new light. Instead, during the city's elaborate infrastructural overhaul associated with the post–World War II period, during which many new expressways, tunnels, bridges, and airports were built, these previous transport plans remained dismissed and forgotten.

Contrasting Visions of Transport Labor

Upon its distribution to the members of the Lusk Committee in 1919, Thomas Tunney's *Throttled!* added to political and business elites' already unsettling image of civil strife in America an equally frightening image of subversive groups of water pilots, factory workers, and garage tinkers commandeering urban industrial and transportative spaces for violent ends. Tunney's memoirs expressed fears of much stolen coal tar, potash, TNT, ammonia, and dynamite being trafficked by subversive freight handlers of obscure origins, affiliations, and loyalties, a seeming hydra of hazily conflated Germans, anarchists, Wobblies, and anti-imperial revolutionaries. Members of the New York financial community who were involved in the Lusk Committee and related anticommunist offensives during the Red Scare of the early 1920s would exert their influence over the *Regional Plan* later that decade. Following a dramatic upswing in the incidence of violence in American class conflict, including a major bombing at Wall Street, such figures felt they had reason to be frightened for their lives. Executing raids, expelling socialist politicians from government, and deporting many immigrants suspected of harboring subversive attitudes formed one type of response to this specter of destruction, though this response had its limits; for instance, a citywide police strike in Boston in 1919 signaled that the manpower necessary to run anti-subversive and anti-criminal raids might not always be readily at hand.[131]

Another type of response to the specter of violent subversion was to rethink the patterns of investment tied up in those spaces around the port noted by Tunney for their "difficulties as areas to be policed." One heavily studied approach to restructuring the Port of New York was already being offered by the 1920 Port and Harbor Development Commission, whose *Comprehensive Plan* was published just months before the Wall Street bombing: invest in the city center as a space where freight was both consumed and produced while also asserting greater top-down control over these freight flows, directing the bulk of urban cargo through a semi-automated freight subway system, to be administered by a centrally run, bistate port authority. Bound up in the Port and Harbor Development Commission's proposal was a set of attitudes toward urban manufacturing labor—attitudes that perceived such labor as valuable, while at the same time seeking new methods for controlling and institutionalizing that labor from above.

To this end, it is perhaps no great surprise that the Port and Harbor Development Commission's lead political counsel was the prominent New York City lawyer Julius Henry Cohen, who had made a name for himself earlier in the 1910s arbitrating on behalf of factory owners during an especially bitter ladies' garment workers strike spearheaded by the socialist organizer Meyer London, in the wake of the horrific inferno at the Triangle Shirtwaist Factory.[132] Cohen had gone on to author some of the pathbreaking American labor arbitration protocols of the day, embraced by Samuel Gompers's craft-unionist American Federation of Labor.[133]

Another approach to rethinking the spaces of the port—and the approach that, for the most part, came to define the harbor's subsequent spatial and transportative development, as well as the city's economic life after World War II—was articulated by the Wall Street–backed Russell Sage Foundation in the 1910s and 1920s, in its two principal publications dealing with the Port of New York: dissolve the city's large corps of freight handlers, whether (as Barnes advocated in his 1915 study) by identifying and dissolving a targeted group, the "shenangoes" or "Chenangoes," within the port's division of labor, or (as the *Regional Plan* advocated during the 1920s) by divesting from freight-handling facilities in the city altogether. This two-pronged approach too was bound up in a distinctive set of attitudes toward labor about the port. In Barnes's telling, the "shenangoes" of New York Harbor were mere drunkards and soft-muscled lowlifes. They offended the concerned early twentieth-century progressivist's eye, perhaps, and brought the rest of the stevedoring profession into ill repute, but did not pose any particular physical threat or danger. In Tunney's telling, by contrast, the "Chenangoes," though romantic in appearance, were implicated in the subversive, clandestine movement of bomb materials around the port. They were multilingual and multinational, anonymous, migrant, fluid, obscure in their origins. They were, in effect, a "hydra," and one whose very name appeared to connote an unsettling local history of harbor bombs and civil warfare; a hydra, then, with TNT and coal tar as its lifeblood, secret bomb-making shops as its organs, and the port's many lighters, tugs, barges, car floats, and freight tracks as its veins. If the *Comprehensive Plan*'s Julius Henry Cohen saw on the docks and in the vessels and factories of New York an urban labor force that could be reasoned with and placated through his juridical innovations in labor arbitration, the Wall Street backers of the *Regional Plan* saw in these spaces something closer to the characterizations in *Throttled!*: a monster to be destroyed.

Conclusion

I have argued that fears of political revolt, of subversive intercourse, and of unrest from below can shape ruling regimes' preferences regarding what sorts of transportation technologies to invest in and what sorts to abandon or destroy. This is not to suggest that other influencing factors, such as the motives of profit and power, are somehow less important to the historical or geographic picture than originally supposed. My aim has been rather to urge the addition of another dimension of social analysis to the broader scholarly conversation regarding the history and development of human mobility systems. The analytical move of focusing on those transport technologies that ruling political regimes have understood as especially useful for insurgents can bear much fruit for the researcher interested in complicating and enriching the broader human narrative of transportation building and unbuilding.

Another area of scholarly conversation to which these findings should prove especially valuable is that focused on surveillance and social legibility as modern political imperatives influencing much top-down decision making regarding social engineering and social rearrangement. Theorists such as James C. Scott have contended that the rise of the modern state, backed by varying degrees of collusion between industrial elites and bureaucratic organizations, has entailed a number of related ruling-class efforts to "make a society legible, to arrange the population in ways that [simplify] the classic state functions of taxation, conscription, and prevention of rebellion."[1] Despite its intrinsic interest in problems of spatial mobility and spatial manipulation, such scholarship has not examined how those imperatives might play out when it comes to the selection or rejection of entire modes or technological systems of transportation. I hope this book has provided the beginnings of such an examination.

James Vance, the same geographer who wrote of the atmosphere of abandonment hanging over the Black Country and Birmingham canal courses during the 1960s—the canals that were "always walled away from the currently used part of the city"—has also written that, in studying the human "transportation revolution" of the modern era, "we must recall the eternal human desire to 'capture the horizon' by reaching it and pushing it ever farther away, through harnessing to human needs the forces of muscle power, of nature, and finally of machine power." In the current epoch, he continues, "our control of the earthly horizon is essentially complete. Every spot on earth is within economic reach for those who live in prosperous and developed countries. No terrestrial economic horizons remain closed to those who have carried the transportation revolution almost to its ultimate fulfillment."[2]

In a sense, this assessment of modern transportation geography is quite accurate. Such widespread, mechanically sophisticated transport technologies as trains, automobiles, and airplanes (to say nothing of technologies for the transfer of digital information) serve in part to close the temporal gap between locations that, in a not so distant past, could be separated by journeys lasting many months or even years. But in another sense, one that I have tried to articulate in this book, such an assessment misses something: many of the same modern political regimes that have managed to "capture" these mechanically sophisticated, time-compressing new capacities for movement across physical space have also, in this same process, *closed off* significant horizons for human transport.

In certain times and places, such closure of transport horizons has been tangled up with a fearful perception that some transport techniques are intrinsically useful for the mobility of rebels and insurgents. Fears of this sort are intimated in the deep social memories and shared habits of associative thinking encapsulated in certain myths and etymologies: the dual meanings of, or mythological backstories to, words like hydra, mulatto, filibuster, Chenango, *demping*, pidgin, and Yangjingbang. Such fears are suggested in descriptive tropes equating the movement of rebel cameleers with the hostile movements of elemental nature, associating canal people with Satanism and treason, mule handlers and canallers with Roma or gypsies, cameleers and canal people with sea pirates, and so on. They are conveyed in concerns surrounding the smuggling abilities of carrier pigeons, canal boatpeople, mule drivers, and elephant mahouts; in descriptions of mule-mounted banditry on the Appalachian backcountry's "Moonshine Front," in Frances Duberly's and George Malleson's accounts of the evasive movements of Tantia Topi's elephant convoy during the Sepoy Mutiny; and in George Perkins Marsh's concerns that importing camels to a desert frontier could serve to increase "the powers of mischief" of that frontier's inhabitants. Equally, they are conveyed in the Earl of Kimberley's warnings, before the British Parliament in 1871, that in Ireland, the canals were "in the hands of the conspirators" (the anti-imperialist Ribbon gang), and also in Thomas Tunney's narrative portrayals of water-borne insurgents carrying infernal devices across New York Harbor during the 1910s.

Such narratives, vignettes, perceptions, encounters, mythologies, and etymologies deserve a broader place in our general understanding of the history of transportation and human mobility—a history that all too often has been told purely in terms of various powerful groups' search for new markets and profits. Indeed, I have contended that the perceptual history treated in this study sheds new historical light on several modern outcomes. These include the disappearance of the mule corps from the U.S. military, British colonialists' hesitance to bring Indian mahouts into East Africa, checks against the importation of camels to aid in the exploration of resource frontiers in the desert, and official efforts to eradicate the Chukchi people's "guerrilla" sled dog. Such analysis also helps place in a more complete social and political context the lack of shipping lanes in the Canadian Midwest and the British interior, the paucity of watercraft in New York Harbor, and the related absence of investment in an inner-city freight-carrying infrastructure for New York City after World War I.

These outcomes were not the result of narrowly impersonal laws of technological or economic out-competition. Rather, many of the qualities that make pigeons, mules, elephants, camels, sled dogs, canal systems, and urban industrial waterfronts useful for insurgent logistics can also make these transportative methods and spaces intrinsically useful for other kinds of high-value activities: rescues by elephant or surficial explorations by camel; in the British canals case, the bulk shipment of grain from Canada to Britain; in the Port of New York case, the flow of needed industrial freight in support of an innovative urban manufacturing cluster. The modern constriction of this field of transportative and social possibilities must be understood in relation to a history of political biases directed against certain less legible classes of transport and transport labor.

In the twenty-first century, we may be living through or approaching the ultimate fulfillment of a transport revolution of some kind. Ours is an age of expansive (if also selective) new forms of connectivity, attended by many sharp political controls over transport and movement: an age of tracked travel, of oligarchically controlled energy for vehicular propulsion, and of communication networks bound by equipment well beyond most of the communicators' reach. Associated with the emergence of this transport geography has been a technological retreat from an older range of possibilities for human mobility, cultural and economic intercourse, and political resistance. Such social and spatial frontiers have become hidden from view. Beyond them, a world remains to be won.

Notes

Introduction

1. Adrien de Gerlache de Gomery, *Belgium in War Time*, trans. Bernard Miall (New York: Doran Co., 1915), 200–201.

2. Severn Valley Railway, "Salvage Collecting in Shrewsbury," in *WW2 People's War Archive Online* (BBC, November 25, 2005), http://www.bbc.co.uk/ww2peopleswar (last accessed Feb. 6, 2013).

3. David Whiteley, "Mill Dovecotes," *British Archaeology* 38 (1998).

4. "Koopman Colony Overwhelmed by Loss … Father Koopman Has Passed Away," *Pigeon Paradise Online*, January 13, 2010, http://www.pipa.be (last accessed Feb. 6, 2013).

5. Malcolm J. Turnbull, "Safe Haven: Records of the Jewish Experience in Australia (Research Guide)," National Archives of Australia, Canberra, 1999. Page numbers not marked. See introduction and chapter 5.

6. Leonard Schroeter, *The Last Exodus* (Seattle: University of Washington Press, 1979), 33.

7. Victoria Schofield, *Afghan Frontier: Feuding and Fighting in Central Asia* (New York: Tauris Parke, 2003), 338.

8. Louise Watt, "China: Pigeons Must Stay in Coops during Congress," *Inquirer News*, Nov. 1, 2012.

9. Che Guevara, *Guerrilla Warfare* (New York: Ocean Press, 2006 [1961]), 36.

10. Andrew Blechman, *Pigeons: The Fascinating Saga of the World's Most Reviled and Revered Bird* (New York: Grove Press, 2007), 38.

11. Gordon Hayes, *The Pigeons That Went to War* (Tulsa: EDC Publishing, 1981), 10.

12. Blechman, *Pigeons*, 39.

13. Erin Fitzgerald, "Cell 'Block' Silence: Why Contraband Cellular Telephone Use in Prisons Warrants Federal Legislation to Allow Jamming Technology," *Wisconsin Law Review*, 2010, 1269.

14. Sanjay Khajuria, "Not Peace Doves?," *Times of India*, May 22, 2010, 12.

15. Hayes, *The Pigeons That Went to War*, 10.

16. See in particular Peter Linebaugh and Marcus Rediker, *The Many-Headed Hydra: Sailors, Slaves, Commoners, and the Hidden History of the Revolutionary Atlantic* (Boston: Beacon Press, 2001), esp. chaps. 5, 6, 7, and the conclusion.

17. See, for instance, William Switala's studies on the geography of the underground railroad in the antebellum American Northeast. These studies can be found in Switala's 2004 *Underground Railroad in Delaware, Maryland, and West Virginia,* his 2006 *Underground Railroad in New Jersey and New York,* and his 2008 *Underground Railroad in Pennsylvania* (all three published by Stackpole Books of Mechanicsburg, PA).

18. See, for example, the work of geographers like Don Mitchell and Timothy Cresswell. Mitchell's work on the subversive mobility of migrant agricultural labor in California during the early twentieth century is to be found in his 1996 *The Lie of the Land: Migrant Workers and the California Landscape* (Minneapolis: University of Minnesota Press), esp. 58–82. Timothy Cresswell's discussion of tramp mobility and radical politics in the United States is in his 2001 *The Tramp in America* (New York: Reaktion Books), esp. chaps. 1, 2, and 7.

19. See, for instance, Andrew Charlesworth, ed., *Atlas of Rural Protest in Britain, 1548–1900* (Philadelphia: University of Pennsylvania Press, 1983), and Andrew Charlesworth, David Gilbert, Adrian Randall, and Humphrey Southall, *An Atlas of Industrial Protest in Britain, 1750–1990* (New York: St. Martin's Press, 1996).

20. James C. Scott, *Seeing Like a State: How Certain Schemes to Improve the Human Condition Have Failed* (New Haven, CT: Yale University Press, 1998), 2. Scott explores (on pp. 53–63) the effects of these sorts of political motivations on the laying out of roads, but not on the selection or rejection of vehicular technologies (that is, modes of transportation).

21. Linebaugh and Rediker, *The Many-Headed Hydra*, 331.

Chapter 1: Mules and Upland Banditry

1. Che Guevara, *Guerrilla Warfare* (New York: Ocean Books, 2006 [1961]), 36.

2. Ibid. Need for tinsmiths and blacksmiths: 117. Shoe factories: 36, 97, 116, 117. Leathersmiths: 97, 117. Saltworks and dryers: 98. Clothing making: 98. Armories: 117. Service batteries and cigarette factories: 118. Camouflage material: 84. "Printing press and a mimeograph stone": 98. Surgical equipment: 63.

3. Ibid., 100.

4. Ibid., 36.

5. Ibid., 100.

6. Ibid., 102.

7. Ibid., 99.

8. George B. Ellenberg, *Mule South to Tractor South: Mules, Machines and the Transformation of the Cotton South* (Tuscaloosa: University of Alabama Press, 2007).

9. Emmett Essin, *Shavetails and Bellsharps: The History of the U.S. Army Mule* (Lincoln: University of Nebraska Press, 2000), 3.

10. Anna Waller, *Horses and Mules and National Defense* (U.S. Office of the Quartermaster General, 1958), 2–31.

11. Melvin Bradley, *The Missouri Mule: His Origin and Times* (Columbia: University of Missouri Press, 1993), 446.

12. Essin, *Shavetails and Bellsharps*, 174.

13. Ibid., 178.

14. Bradley, *The Missouri Mule*, 437.

15. Michael Parrino, *An Introduction to Pack Transport and Pack Artillery* (New York: Queensland Publishing, 1956), 36–88.

16. Ibid., 86.

17. Jack Raymond, "Army to Retire Its Combat Mules," *New York Times,* Dec. 2, 1956, 1.

18. Essin, *Shavetails and Bellsharps*, 159, and see Richard Bulliet, *The Camel and the Wheel* (Cambridge, MA: Harvard University Press, 1975), 255.

19. See in particular Parinno, *An Introduction to Pack Transport and Pack Artillery,* and Waller, *Horses and Mules and National Defense.*

20. Special Warfare Agency, *Final Study: U.S. Army Requirement for Pack Animals* (U.S. Army Combat Development Command, Special Warfare Agency, 1965), 16.

21. Bradley, *The Missouri Mule*, 404, and K. I. Barlow, "Dropping of Mules by Parachute," *Journal of the Royal Army Veterinary Corp* 17 (1946): 93–98.

22. Bradley, *The Missouri Mule*, 433.

23. See discussion in Charles Shrader, *The Withered Vine: Logistics and the Communist Insurgency in Greece, 1945–1949* (Westport, CT: Greenwood Publishing Group, 1999), esp. 3 and 144. On 144: "Insofar as equipment and supplies provided by the GDA's

outside supporters reached these forces, they did so by ground transport—primarily pack mule convoys and porters, many of whom were women.... Although forced to operate in rugged, unimproved areas, the GDA had not the manpower, engineering skill or equipment to do much more than improve the mountain trails, erect temporary bridging, and try to keep the entire network in a basic state of repair."

24. Col. J. C. Murray, "Anti-bandit War II," *Marine Corps Gazette* 38 (February 1954): 55.

25. Waller, *Horses and Mules and National Defense*, 31; Essin, *Shavetails and Bellsharps*, 194–195.

26. Essin, *Shavetails and Bellsharps*, 196.

27. Tom Wicker, "Gee-Haw for the Army," *New York Times,* Apr. 12, 1985, A27.

28. Essin, *Shavetails and Bellsharps*, 202.

29. Tony Perry, "Marines' Beasts of Burden Again Leading the Pack," *Los Angeles Times,* July 7, 2009. And see Susan Orlean, "Mules in the Modern Military," *New Yorker,* Feb. 15, 2010.

30. See, for instance, Arnulf Grübler, *The Rise and Fall of Infrastructures* (Heidelberg: Physica-Verlag, 1990), 268.

31. Dwight Eisenhower, "Farewell Address" (1961), in *America in the 20th Century XII* (Tarrytown, NY: Marshall Cavendish, 2003), 1622.

32. Robert Lamb, *The Mule in Southern Agriculture* (Berkeley: University of California Press, 1963), 17–22.

33. Melvin Bradley, *The Mule Industry of Missouri Remembered* (Columbia: University of Missouri Press, 1991), 18.

34. Bradley, *The Missouri Mule*, 448; Ellenberg, *Mule South to Tractor South*, 53.

35. "Another Army Test," *Motor Age,* Apr. 11, 1912, 37.

36. See "Gasoline May Banish Army Mule," *Georgian and News,* Mar. 11, 1912, and "Army Mule Banished," *Schenectady Gazette,* Mar. 12, 1920, 4.

37. Essin, *Shavetails and Bellsharps*, 143–144. See also Luz María Hernández Sáenz, "Smuggling for the Revolution: Illegal Traffic of Arms on the Arizona-Sonora Border, 1912–1914," *Arizona and the West* 28, no. 4 (1986): 357–377.

38. Essin, *Shavetails and Bellsharps*, 156.

39. Bradley, *The Missouri Mule*, 210.

40. For an overview of the historical debate over the reason for the discrepancy, in the United States, between the southern preference for mules and the northern preference for horses, see Martin Garrett, "Mules in Southern Agriculture: Revisited,"

Journal of Agricultural and Applied Economics 33, no. 3 (2001): 583–590. One rather odd bias that has characterized this decades-old debate (including Garrett's article) is the tendency of interested scholars to treat the southern preference for mules as the puzzle to be solved, rather than the northern *aversion* to mules.

41. Garrett, "Mules in Southern Agriculture," 587.

42. Bradley, *The Missouri Mule*, 146.

43. Lamb, *The Mule in Southern Agriculture*, 26.

44. Bradley, *The Missouri Mule*, 18, 146.

45. William Ferris, *Mule Trader: Ray Lum's Tales of Horses, Mules and Men* (Oxford: University of Mississippi Press, 1998), 29, 108.

46. Marguerite Riordan, "The Irish Mule Traders," *American Cattle Producer,* October 1950, 9–28. See also Ferris, *Mule Trader*, 108 and 231.

47. Bonny Ball and Randy Hodge, *Melungeon: Notes on the Origins of a Race* (Johnson City, TN: Overmountain Press, 1992), 74.

48. Jean-Jacques Rousseau, *Confessions*, trans. S. W. Orson (London: Aldus Society, 1903 [1782]), 592–593.

49. Teresa Reed, "Sometimes I Let It Age Ten or Fifteen Minutes," *Bittersweet* 3, no. 1 (1975): 55.

50. Joseph Earl Dabney, *Mountain Spirits: A Chronicle of Corn Whiskey from King James' Ulster Plantation to America's Appalachians and the Moonshine Life* (New York: Charles Scribner's Sons, 1974), 186.

51. Ibid., xxvi, 13.

52. Clark B. Firestone, *Bubbling Waters* (New York: Robert McBride and Co., 1938), 139.

53. Dabney, *Mountain Spirits*, 125.

54. Ibid., 209.

55. Ibid., 4.

56. David Corbin, *Life, Work and Rebellion in the Coalfields: The Southern West Virginia Miners, 1880–1922* (Urbana: University of Illinois Press, 1989), 149.

57. Ibid., 37, 56.

58. Ibid., 130, 220.

59. See Lon Savage, *Thunder in the Mountains: The West Virginia Mine War* (Pittsburgh: University of Pittsburgh Press, 1990), and Michael Meador, "The Red Neck War of 1921," *Goldenseal*, April–June 1981, 44–57.

60. Meador, "The Red Neck War of 1921," 56; Christopher Swope, "The Battles of Blair Mountain," *Preservation,* May/June 2006.

61. Meador, "The Red Neck War of 1921," 49; Clayton Laurie, "The United States Army and the Return to Normalcy in Labor Dispute Interventions: The Case of the West Virginia Coal Mine Wars, 1920–1921," *West Virginia History* 50 (1991): 1–24; Robert Shogun, *Battle of Blair Mountain: The Story of America's Largest Labor Uprising* (Boulder, CO: Basic Books, 2006), 200–205.

62. Essin, *Shavetails and Bellsharps,* 143–144.

63. William Shephard, "The Hootch Runners," *Cosmopolitan,* January 1922, 117.

64. "In Long Hunt for Treasure in the Southwest," *Chronicle-Telegram of Elyria,* Oct. 12, 1928.

65. Haldeen Braddy, "Smugglers' Argot in the Southwest," *American Speech* 31, no. 2 (1956): 96–101.

Chapter 2: Transportation across Intermediate States of Matter

1. Shelby Tucker, *Among Insurgents: Walking through Burma* (London: Radcliffe Press, 2000), 106–112.

2. Ibid., xvi.

3. Ibid., 166.

4. Ibid., xvi, 110.

5. Ibid., 198–199.

6. Ibid., 264.

7. Major A. W. Baird[?], "Kathiawar," *Gazetteer of the Bombay Presidency* 8 (1884): 56.

8. Ibid., 57.

9. C.W.S. Kinnersley, "Notes of a Tour through the Siamese States," *Journal of the Straits Branch of the Royal Asiatic Society,* July 1901, 64.

10. S. Barkataki, *Assam* (New Delhi: National Book Trust, 1969), 119.

11. Nigel Pankhurst, "'Elephant Man' Who Staged Daring WWII Rescues," *BBC News,* Oct. 31, 2010. See also Geoffrey Tyson, *Forgotten Frontier* (Calcutta: W. H. Targett, 1945), 99–120. And see Andrew Martin, *Flight by Elephant: The Untold Story of World War Two's Most Daring Jungle Rescue* (London: Fourth Estate, 2013).

12. Gyles Mackrell, *Chaukan Operations,* Film footage, 1942, available at the British Film Institute, London.

13. Jayantha Jayewardene, *The Elephant in Sri Lanka* (Colombo: Wildlife Heritage Trust of Sri Lanka, 1994), 25.

14. Tucker, *Among Insurgents*, 236.

15. Dhriti K. Lahiri-Choudhury, *The Great Indian Elephant Book: An Anthology of Writings on Elephants in the Raj* (Oxford: Oxford University Press, 1999), 147.

16. Ibid., 150.

17. Charles Eden, *India, Historical and Descriptive* (London: Marcus Ward, 1876), 198–200.

18. Richard Lair, *Gone Astray: The Care and Management of the Asian Elephant in Domesticity* (Bangkok: FAO Office for Asia and the Pacific, 1997), 127. See also Khim Zaw, "Capturing and Training of Wild Elephants for Utilization in Timber Harvesting," *Myanmar Forestry Journal* 4, no. 3 (2000): 8–14, 12.

19. Richard Hughes, "Gems and Junkies in Burma," *The Guide (The Gem Guide)* 20, no. 4(5) (2001): 8–14.

20. Lair, *Gone Astray*, 199.

21. Raman Sukumar, *The Asian Elephant: Ecology and Management* (Cambridge: Cambridge University Press, 1993), 37; Robert Olivier, "Distribution and Status of the Asian Elephant," *Oryx* 14 (1978): 379–424, 404; Ronald Nowak, "Wildlife of Indochina: Tragedy or Opportunity?," *National Parks and Conservation Magazine*, June 1976, 13–18, 18; John Prados, *The Blood Road: The Ho Chi Minh Trail and the Vietnam War* (New York: John Wiley, 1999), 221; John Prados and Ray Stubbe, *Valley of Decision: The Siege of Khe Sanh* (New York: Houghton Mifflin, 1991), 66.

22. Bertil Lintner, *Land of Jade: A Journey from India through Northern Burma to China* (Bangkok: Orchid Press, 1989), 179.

23. Jack Fong, *Revolution as Development: The Karen Self-Determination Struggle against Ethnocracy* (Boca Raton, FL: Universal Publishers, 1994), 142. Jonathan Falla, in *True Love and Bartholomew: Rebels on the Burmese Border* (Cambridge: Cambridge University Press, 1991), also talks about Karen rebel mahouts, especially on pages 128–134.

24. George Malleson, *History of the Indian Mutiny: Commencing from the Close of the Second Volume of Sir John Kaye's History of the Sepoy War*, vol. 3 (London: Longmans, Green and Co., 1896 [1878]), 308.

25. Ibid., 319–320.

26. Quoted in Frances Isabella Duberly, *Campaigning Experiences in Rajpootana and Central India during the Suppression of the Mutiny, 1857–1858* (London: Smith, Elder and Co., 1859), 163.

27. Ibid., 166.

28. Ibid., 16.

29. Malleson, *History of the Indian Mutiny*, 321.

30. Ibid., 324–326.

31. Ibid., 329.

32. Duberly, *Campaigning Experiences in Rajpootana*, 198.

33. Ibid., 202.

34. Malleson, *History of the Indian Mutiny*, 329–330.

35. Duberly, *Campaigning Experiences in Rajpootana*, 185.

36. Ibid., 200.

37. Ibid., 201.

38. Construction on the Gandhi Sagar Dam on the Chambal began in 1953. The Rajghat Dam on the Betwa was begun in 1975.

39. Dwarika Dhungel and Santa Pun, *The Nepal-India Water Relationship: Challenges* (Heidelberg: Springer, 2009), 70. Not to be confused with the East Rapti River, which drains the Chitwan Valley.

40. William Howitt, *Cassell's Illustrated History of England,* vol. 8 (London: Cassell, Petter and Galpin, 1864), 517.

41. Charles Ball, *The History of the Indian Mutiny: Giving a Detailed Account of the Sepoy Insurrection in India: and a Concise History of the Great Military Events which Have Tended to Consolidate British Empire in Hindostan* (London: London Print and Publishing Co., 1858), 555.

42. Jules Verne, *The Steam House* (New York: Charles Scribner's Sons, 1881), 171.

43. Eden, *India, Historical and Descriptive*, 262–264.

44. John Harris, *The Indian Mutiny* (Ware, UK: Wordsworth, 2001), 79–80; Pratul Chandra Gupta, *Nana Sahib and the Rising at Cawnpore* (Oxford: Clarendon Press of Oxford University Press, 1963), 112.

45. Malleson, *History of the Indian Mutiny*, 308–310.

46. G. Schweinfurth, F. Ratzel, R. W. Felkin, and G. Hardlaub, eds., *Emin Pasha in Central Africa: Being a Collection of His Letters and Journals* (London: George Philip and Son., 1888), 390.

47. Tucker, *Among Insurgents*, 217.

48. Ibid., 204.

49. Ibid., 236–237.

50. Duberly, *Campaigning Experiences in Rajpootana*, 244.

51. Hope Werness, *Encyclopedia of Animal Symbolism in Art* (New York: Continuum, 2006), 160.

52. Malleson, *History of the Indian Mutiny*, 318.

53. Quoted in Duberly, *Campaigning Experiences in Rajpootana*, 227.

54. Ellen Semple, "Pirate Coasts of the Mediterranean Sea," *Geographical Review* 2 (1916): 134–151, 138.

55. T. E. Lawrence, *The Seven Pillars of Wisdom: A Triumph* (Garden City, NY: Doubleday, 1966 [1926]), 8–9.

56. Ibid., 208.

57. T. E. Lawrence, "Guerrilla," in *Encyclopaedia Britannica* (London: Encyclopedia Britannica Co., 1932), 952.

58. Ned Kahn, "Turbulent Architecture," lecture, Hendricks Chapel, Syracuse University, Syracuse, NY, Oct. 5, 2006.

59. R. A. Bagnold, *The Physics of Blown Sand and Desert Dunes* (Mineola, NY: Courier Dover, 2005 [1941]), 219.

60. Robert Christopherson, *Geosystems: An Introduction to Physical Geography*, 7th ed. (Upper Saddle River, NJ: Pearson Preston Hall, 2009), 469–475.

61. Michael Welland, *Sand: The Never-Ending Story* (Berkeley: University of California Press, 2009), 164.

62. Ibid., plate 11.

63. Richard Bulliet, *The Camel and the Wheel* (Cambridge, MA: Harvard University Press, 1975), 247.

64. Ibid., 253.

65. Ibid., 254.

66. George Perkins Marsh, *The Camel: His Organization, Habits and Uses* (Boston: Gould and Lincoln, 1854), 188–189.

67. Ibid. On pages 21, 75–78 (where there is a lengthy quote), and 93, Marsh cites from Abdelkader's notes in Abdelkader and Eugène Daumas's 1851 *Le Chevaus du Sahara* (Paris: F. Chamerot).

68. Percy Cross Standing, *Guerrilla Leaders of the World* (Boston: Houghton Mifflin, 1913), 106, and see Raphael Danzinger, *Abd al-Qadir and the Algerians: Resistance to the French and Internal Consolidation* (New York: Holmes and Meier, 1997), 120.

69. Charles Callwell, *Small Wars: Their Principles and Practice* (London: His Majesty's Stationery Office, 1903), 110.

70. H. M. Barker, *Camels and the Outback* (Melbourne: Sir Isaac Pitman and Sons, 1964), 88–89.

71. Bulliet, *The Camel and the Wheel*, 253.

72. Philip Jones and Anna Kenny, *Australia's Muslim Cameleers: Pioneers of the Inland* (Adelaide: Wakefield Press, 2007), 10.

73. Barker, *Camels and the Outback*, 91.

74. Ibid., 92.

75. It may be worth noting, though, that police would regularly raid the "Ghan-towns" to seize unlicensed revolvers. See Christine Stevens, *Tin Mosques and Ghan-towns: A History of Afghan Cameldrivers in Australia* (Oxford: Oxford University Press, 1989), 254.

76. Walter Truscott, "Recollections of Suakim," *English Illustrated Magazine* 6 (1889): 714.

77. Standing, *Guerrilla Leaders of the World*, 249, 253.

78. Ibid., 249.

79. Truscott, "Recollections of Suakim," 707.

80. Douglas Pike and Geoffrey Serle, *Australian Dictionary of Biography* (Melbourne: Melbourne University Press, 1988), 272.

81. Bureau of Agriculture of Western Australia, *The Journal of the Bureau of Agriculture* 4 (1897): 1267.

82. Western Australia Parliament, *Parliamentary Debates: Legislative Council and Legislative Assembly* 19 (1902): 928.

83. Jones and Kenny, *Australia's Muslim Cameleers*, 19.

84. Barker, *Camels and the Outback*, 93.

85. D. Appleton, *Annual Cyclopaedia and Register of Important Events,* vol. 5 (New York: Appleton, 1885), 294.

86. Ibid., 295–296.

87. Robert Brown, *The Story of Africa and Its Explorers,* vol. 2 (London: Cassell, 1893), 67; Appleton, *Annual Cyclopaedia*, 296.

88. See "Expéditions Belges Africains," *Revue Géographique International* 6 (1881): 81; Appleton, *Annual Cyclopaedia*, 296; and also L. K. Rankin, "The Elephant Experiment in Africa," *Proceedings of the Royal Geographic Society* 4 (1882): 273–289. For the

likely location of the Tabora research station, see "Simba (Simbo)," in Herbert Friedmann and Arthur Loveridge, "Notes on the Ornithology of Tropical East Africa," *Bulletin of the Museum of Comparative Zoology at Harvard College* 81, no. 1 (1937): 21.

89. Trevenen Holland and Henry Hozier, *Record of the Expedition to Abyssinia, Vol. II* (London: Her Majesty's Stationary Office, 1870), 86, 230.

90. "The Proposed Overland African Telegraph," *Proceedings of the Royal Geographical Society* 1 (1879): 268.

91. "Letter to Professor G. Schweinfurth," dated Lado, Dec. 25, 1881, in G. Schweinfurth et al., *Emin Pasha in Central Africa*, 390.

92. Colonial Office of Great Britain, *Correspondence Relating to the Preservation of Wild Animals in Africa, London* (London: His Majesty's Stationery Office, 1906), 187. Includes "Letter from Lieutenant-Colonel Delme Radcliffe to Commissioner Consul-General Sadler, Uganda, of the Anglo-German Boundary Commission, Entebbe."

93. Hesketh Bell, *Glimpses of a Governor's Life, from Diaries, Letters and Memoranda* (London: Sampson, Low, Marston and Co., 1946), 197–200.

94. Tucker, *Among Insurgents*, 166.

95. Alastair Scobie, *Animal Heaven* (London: Cassell and Co., 1953), 142–143.

96. Lair, *Gone Astray*, 5.

97. Armand Denis, *On Safari: The Story of My Life* (New York: E. P. Dutton, 1963), 85.

98. Murray Fowler and Susan Mikota, *Biology, Medicine, and Surgery of Elephants* (Ames, IA: Wiley-Blackwell, 2006), 19.

99. Peter Linebaugh and Marcus Rediker, *The Many-Headed Hydra: Sailors, Slaves, Commoners, and the Hidden History of the Revolutionary Atlantic* (Boston: Beacon Press, 2001), 2.

100. Ibid., 2–3.

101. Ibid., 173.

102. William Carey's 1827 *Dictionary of the Bengalee Language* (Serampore, India, 327) translates *naga* (नाग) thus: "in Hindoo mythology a race of fabulous beings; the spectacle snake (Coluber Naga); a hydra, a serpent, an elephant, a species of grass (Cyperus pertenuis)." See also Jean Philippe Vogel, *Indian Serpent-lore; or, The Nagas in Hindu Legend and Art* (London: A. Probsthain, 1926), 129. And see Werness, *Encyclopedia of Animal Symbolism in Art*, 160, 267.

103. Linebaugh and Rediker, *The Many-Headed Hydra*, 179.

104. Ibid., 182–183.

105. Lorna Demidoff and Michael Jennings, *The Complete Siberian Husky* (New York: Howell Book House, 1978), 42.

106. Benedict Allen, "An Iceman's Best Friend," *Geographical Magazine* 78 (December 2006).

107. F. Berkes and D. Jolly, "Adapting to Climate Change: Social-ecological Resilience in a Canadian Western Arctic Community," *Conservation Ecology* 5, no. 2 (2001): 10.

108. Richard Sale and Eugene Potapov, *Scramble for the Arctic: Ownership, Exploitation and Conflict in the Far North* (London: Frances Lincoln, 2010), 58.

109. Anna Reid, *The Shaman's Coat: A Native History of Siberia* (New York: Walker and Co., 2003), 178.

110. Demidoff and Jennings, *The Complete Siberian Husky*, 42.

111. Allen, "An Iceman's Best Friend."

112. Waldemar Bogoras, *The Chukchee* (New York: G. E. Stechart and Co., 1904), 652.

113. Reid, *The Shaman's Coat*, 184.

114. Demidoff and Jennings, *The Complete Siberian Husky*, 48.

115. Ibid.

116. Ibid., 40.

117. James Forsyth, *A History of the Peoples of Siberia* (Cambridge: Cambridge University Press, 1992), 150.

118. Demidoff and Jennings, *The Complete Siberian Husky*, 53.

119. D. Ledovsky, "A Dog's Life: Start and Finish," *New Times International: A Russian Weekly* 22 (1992): 32.

120. Demidoff and Jennings, *The Complete Siberian Husky*, 50.

121. Ledovsky, "A Dog's Life," 30.

122. Demidoff and Jennings, *The Complete Siberian Husky*, 41, 52–53. The Soviet destruction of the Chukchi sled dog may be comparable to the mass killing of Inuit sled dogs that took place in northern Canada during the 1950s and 1960s. As of 2013, this mass killing is still under investigation. See Qikiqtani Truth Commission, *Thematic Reports and Special Studies, 1950–1975* (Iqaluit, NU: Inhabit Media, 2013).

123. Stephen Wurm, Peter Mühlhäusler, and Darrell Tyron, eds., *Atlas of Languages of Intercultural Communication in the Pacific, Asia, and the Americas* (Berlin: Mouton de Gruyter, 1996), 984.

124. Linebaugh and Rediker, *The Many-Headed Hydra*, 152–154.

125. Ibid., 311.

126. Lareto Todd, *Pidgins and Creoles* (London: Routledge, 1974), 20–22.

127. Ellen Semple, *Influences of Geographic Environment: On the Basis of Ratzel's System of Anthropo-geography* (New York: H. Holt and Co., 1911), 276.

128. Jia Jane Si, *The Genealogy of Dictionaries: Producers, Literary Audiences and the Circulation of English Texts in the Treaty Port of Shanghai* (Philadelphia: University of Pennsylvania, Department of East Asian Languages and Civilizations, 2005), 11, 35.

129. Peter G. Rowe and Seng Kuan, *Shanghai: Architecture and Urbanism for Modern China* (New York: Prestel, 2004), 39.

130. John DeFrancis, ed., *ABC Chinese-English Comprehensive Dictionary* (Honolulu: University of Hawaii Press, 2003), 1110. And see "Buzzwords and Talk Shanghai," *Shanghai Daily*, June 26, 2010, http://www.shanghaidaily.com (last accessed Feb. 6, 2013).

131. Charles Taylor, *Five Years in China: With Some Account of the Great Rebellion, and a Description of St. Helena* (New York: Derby and Jackson, 1860), 80.

132. J. W. Maclellan, *The Story of Shanghai, From the Opening of the Port to Foreign Trade* (Shanghai: North China Herald Office, 1889), 31.

133. Zhang Zhen, *Amorous History of the Silver Screen: Shanghai Cinema* (Chicago: University of Chicago Press, 2006), 47.

134. Letter from Jierhanga to Robert McLane, dated 1856, in Senate of the United States, *The Executive Documents Printed by Order of the Senate of the United States, Second Session, 35th Congress* (Washington, DC: William A. Harris Publisher, 1859), 690. In this document the geographic names Fujian, Guangdong, and Yangjingbang are transcribed as "Fuh-kien," "Kwang-tung," and "Yang-king-pang," respectively. I have inserted the more recent transcriptions for the sake of clarity. "Lin-Lee-chuen" may be a faulty transcription of Liu Lichuan, the leader of the Small Swords occupation of Shanghai during 1853.

135. Jeng Gao James, *Historical Dictionary of Modern China* (Lanham, MD: Scarecrow Press, 2009), 331.

136. Herbert Warington Smyth, *Five Years in Siam: From 1891 to 1896* (London: John Murray, 1898), 108.

137. Charles Shrader, *The Withered Vine: Logistics and the Communist Insurgency in Greece, 1945–1949* (Westport, CT: Greenwood Publishing Group, 1999), 143–145.

138. Tucker, *Among Insurgents*, 343–344.

139. Ibid., 344.

140. John Davenport and Julia Davenport, eds., *The Ecology of Transportation: Managing Mobility for the Environment* (Dordrecht, The Netherlands: Springer Science, 2006), 348.

141. Todd Kemper and Ellen Macdonald, "Directional Change in Upland Tundra Plant Communities 20–30 Years after Seismic Exploration in the Canadian Low-Arctic," *Journal of Vegetation Science* 20 (2009): 557–567.

142. Michael Raffaeli, "Sometimes Dog Mushing Can Be a Drag," *Denali National Park and Preserve: Runnin' with the Kennels,* Jan. 24, 2012, http://www.nps.gov/dena (last accessed Feb. 6, 2013).

Chapter 3: Fly-Boaters, Filibusters, and Canals

1. James Vance, *Capturing the Horizon: The Historical Geography of Transportation since the Transportation Revolution of the Sixteenth Century* (New York: Harper and Row, 1986), 101.

2. Richard Dean, *Canals of Birmingham: No. 2 in the Historical Canal Map Series*, 3rd ed. (Stoke-on-Trent, UK: Cartographics, 2008).

3. Charles Hadfield, *The Canal Age* (Newton Abbot, UK: David and Charles, 1981), 162.

4. See map in European Commission, *Inland Waterway Transport* (Brussels: European Commission, Directorate-General for Energy and Transport, 2002), 2.

5. Hadfield, *The Canal Age*, 69.

6. Philip Bagwell, *The Transport Revolution from 1770* (London: Batsford Press, 1974), 154.

7. Ibid., 154.

8. Hadfield, *The Canal Age*, 162.

9. Royal Commission, *Report of the Royal Commission on Canals and Waterways,*vol. 7 (London: Her Majesty's Stationery Office, 1909).

10. Herbert Quick, *American Inland Waterways* (New York: G. P. Putnam's Sons, 1909), 9.

11. Royal Commission, *Report of the Royal Commission on Canals and Waterways,* vol. 6 (London: Her Majesty's Stationery Office, 1909), 89.

12. J. S. Nettleford, *Garden Cities and Canals* (London: St. Catherine Press, 1914), 133.

13. Ibid., 136.

14. Royal Commission, *Report of the Royal Commission on Canals and Waterways,* vol. 7, 121–125.

15. Ibid., 93–95.

16. Royal Commisssion, *Report of the Royal Commission on Canals and Waterways,* vol. 6, maps 1–3.

17. Nettleford, *Garden Cities and Canals,* 103.

18. Bagwell, *The Transport Revolution from 1770,* 154.

19. Ibid., 271.

20. "A Channel Across Ireland," *Review of Reviews* 9, no. 2 (1894): 134; see also *Van Nostrand's Engineering Magazine* 31, no. 4 (1884): 345; and see Edward Watkin, "Letter from E. Watkin MP to Select Committee of Irish House of Commons on Irish Industries," *Times of London,* June 27, 1885, 9.

21. Mark Twain, *Life on the Mississippi* (Boston: J. R. Osgood, 1883), 398.

22. See Florence Dorsey, *Road to the Sea: The Story of James B. Eads and the Mississippi River* (New York: Rinehart, 1947).

23. Patrick Flanagan, *Transport in Ireland* (Dublin: Transport Research Associates, 1969), 42.

24. Robert Legget, *Canals of Canada* (Newton Abbot, UK: David and Charles, 1976), 23.

25. Ibid., 24.

26. Philip A. Buckner, "1870s Political Integration," in *The Atlantic Provinces in Confederation,* ed. E. R. Forbes and Delphin Muise (Toronto: University of Toronto Press, 1993), 49.

27. Donald J. Savoie, *Visiting Grandchildren: Economic Development in the Maritimes* (Toronto: University of Toronto Press, 2006), 26. And see Marc Shell, *Grand Manan: A Large History of a Small Island* (Montreal: McGill University Press, 2015), chap. 2.

28. Legget, *Canals of Canada,* 25.

29. Charles Hadfield, *World Canals* (Newton Abbot, UK: David and Charles, 1986), 352–355.

30. Shell, *Grand Manan,* chap. 2.

31. Canada Department of Public Works, *Georgian Bay Ship Canal* (Ottawa: C. H. Parmelee, 1908), plate 2.

32. Quick, *American Inland Waterways,* 18.

33. J. A. Latcha, "Railroads versus Canals," *North American Review* 166, no. 495 (1898): 212.

34. "The Ottawa River Canal," *New York Times,* July 16, 1895.

35. Roy Wolfe, "Transportation and Politics: The Example of Canada," *Annals of the Association of American Geographers* 52, no. 2 (1962): 176–190.

36. J. P. Usher, "Letter ... in relation to uniting the navigable waters of the Mississippi River with the Red River of the North by slackwater and canal navigation," in *37th Congress, 3rd Session,* S. M. Doc. 8 (Jan. 5, 1863). The U.S. Congress reexamined the idea in 1894; see J. T. McCleary, "Connecting the Minnesota River and the Red River of the North," in *53rd Congress, 2d Session,* Jl. 27, H. R. 1335 (1894).

37. "Waterway of the North," *New York Times,* Nov. 5, 1895.

38. Lawrence Burpee, "How Canada Is Solving her Transportation Problem," *Popular Science,* September 1905, 455–464, 463.

39. Quick, *American Inland Waterways,* 23–25.

40. Watson Griffin, "Canada: The Land of Waterways," *Bulletin: American Geographical Society of New York,* 1890, 416–417.

41. Ibid.

42. Marjorie Forrester, "Markers on the Forty-Ninth," *Manitoba Historical Society Transactions,* ser. 3 (1959–1960).

43. "Canal Feasible," *Victoria Daily Colonist,* Apr. 6, 1910, 1.

44. Col. W. F. Baker, "Inland Waterways," in *Official Proceedings of the 19th Session of the Trans-Mississippi Congress* (Trans-Mississippi Commercial Congress, 1909), 107.

45. Farley Mowat, *The Dog Who Wouldn't Be* (Thorndike, ME: Thorndike Press, 1957), 3.

46. Ibid., 89.

47. Ibid., 101–102.

48. Ibid., 99.

49. Ibid., 101.

50. Ibid., 106.

51. Exceptions tend to date from the very period under consideration, 1870–1920. See, for instance, Erwin Pratt's 1912 *History of Inland Transportation and Communication in England* (New York: E. P. Dutton), or Harold Moulton's 1912 *Waterways versus Railroads* (New York: Houghton Mifflin).

52. Robert Fogel, *Railroads and American Economic Growth* (Baltimore: Johns Hopkins University Press, 1964), 218.

53. Ibid., 224.

54. Ibid., 235–236.

55. Brian Mitchell, David Chambers, and Nicholas Crafts, "How Good was the Profitability of British Railways, 1870–1912?," *Warwick Economic Research Papers* 859 (2008).

56. Ann M. Carlos and Frank Lewis, "The Profitability of Early Canadian Railroads," in *Strategic Factors in 19th Century American Economic History*, ed. Claudia Goldin (Chicago: University of Chicago Press, 1992), 401–426. See esp. 422.

57. Fogel, *Railroads and American Economic Growth*, 94.

58. Thomas Holmes and James Schmitz, "Competition at Work: Railroads vs. Monopoly in the US Shipping Industry," *Quarterly Review*, Spring 2001, 3–29.

59. Wolfe, "Transportation and Politics," 182.

60. Hadfield, *World Canals*, 343.

61. Vance, *Capturing the Horizon*, 101.

62. Philip Bagwell and Peter Lyth, *Transport in Britain 1750–2000: From Canal Lock to Gridlock* (London: Hambledon and London, 2002), 19.

63. Ibid., 20.

64. Quick, *American Inland Waterways*, 65–66.

65. Deepak Nayyar, "Globalization, History and Development: A Tale of Two Centuries," *Cambridge Journal of Economics* 30 (2006): 140.

66. Wolfe, "Transportation and Politics," 178.

67. Gerry van Houten, *Corporate Canada: An Historical Outline* (Toronto: Progress Books, 1991), 42–43.

68. Nayyar, "Globalization, History and Development," 145.

69. Permanent International Association of Navigation Congresses (1908), *Report of the Proceedings, Eleventh International Navigation Congress, St. Petersburg* (Brussels: Imprimerie des Travaux Publics), 402.

70. Hadfield, *World Canals*, 175–176.

71. Ibid., 170.

72. Data on capital called are from Irving Stone, *The Global Export of Capital from Great Britain, 1865–1914: A Statistical Survey* (Basingstoke, UK: Palgrave Macmillan, 1999).

73. Ron Jones, *The Albert Dock, Liverpool* (Emsworth, UK: R. J. Associates Ltd., 2004).

74. London Dockland Development Corporation, *Royal Docks* (London: LDDC Completion Booklets, 1998).

75. Quoted in Bagwell, *The Transport Revolution from 1770*, 144.

76. W. T. Jackman, *The Development of Transportation in Modern England* (Cambridge: Cambridge University Press, 1916), 658.

77. Ibid., 659.

78. Ernest Oldmeadow, *Coggin* (New York: Century Co., 1920), ix.

79. Ibid., xiii.

80. Ibid., 88.

81. Ibid., 91–93.

82. Charles Hadfield, *British Canals,* 8th ed. (Stroud, UK: Alan Sutton Publishing, 1998), 235.

83. Harry Hanson, *The Canal Boatmen, 1760–1914* (Manchester: Manchester University Press, 1975), 64.

84. Ibid., 64.

85. "Sunday Trading on Canals," *Sessional Papers of the House of Lords*, 4th and 5th Victoria, vol. 21, no. 16 (1841).

86. *First Report of the Commissioners Appointed to Inquire as to the Best Means of Establishing an Effective Constabulary Force of England and Wales* (London: Her Majesty's Stationery Office, 1839), 102.

87. Ibid., 93.

88. Hanson, *The Canal Boatmen*, 67–80.

89. "Sunday Trading on Canals," 81.

90. Ibid., 64.

91. Ibid., 89.

92. Brian Vaughton, "Birmingham Ballads" (BBC Radio, 1962), http://www.cpatrust.org.uk/Bham_ballads.htm (last accessed Feb. 6, 2013).

93. Stella Davis, *Living through the Industrial Revolution* (London: Routledge Press, 1966), 52.

94. Hanson, *The Canal Boatmen*, 1.

95. Ibid., 166.

96. "The Waterways of New York," *Harper's Magazine* 48 (1873): 2.

97. Hanson, *The Canal Boatmen*, 168.

98. Ibid., 48–49.

99. Roy M. MacLeod, "Social Policy and the 'Floating Population': The Administration of the Canal Boats Acts, 1877–1899," *Past & Present* 35 (1966): 102–132, 105.

100. See Assa Doron, *Caste, Occupation and Politics on the Ganges* (Burlington, VT: Ashgate, 2008), 39.

101. Ibid., 42.

102. John Harris, *The Indian Mutiny* (Ware, UK: Wordsworth, 2001), 79–80; Pratul Chandra Gupta, *Nana Sahib and the Rising at Cawnpore* (Oxford: Clarendon Press of Oxford University Press, 1963), 112.

103. Doron, *Caste, Occupation and Politics on the Ganges*, 40.

104. Smita Tewary Jassal, "Caste and the Colonial State: Mallahs in the Census," *Contributions to Indian Sociology* 35 (2001): 319–354, 340.

105. Doron, *Caste, Occupation and Politics on the Ganges*, 40–41. See also Nita Kumar, *The Artisans of Banaras: Popular Culture and Identity* (Princeton, NJ: Princeton University Press, 1995), 126–131.

106. Hadfield, *British Canals*, 236.

107. *Times of London,* June 4, 1877, quoted in Macleod, "Social Policy and the 'Floating Population,'" 111.

108. George Smith of Coalville, *Our Canal Population* (Wakefield, UK: E. P. Publishing, 1875).

109. Macleod, "Social Policy and the 'Floating Population,'" 107–108.

110. Ibid., 110.

111. Mark Guy Pearse, *Rob Rat: A Story of Barge Life* (London: Wesleyan Conference Office, 1878), 23.

112. Ibid., 12.

113. George Smith of Coalville, *Canal Adventures by Moonlight* (London: Hodder and Stoughton, 1881), 68.

114. Ibid., 112.

115. Quoted in Macleod, "Social Policy and the 'Floating Population,'" 113.

116. Oldmeadow, *Coggin*, xiv.

117. Bagwell, *The Transport Revolution from 1770*, 156; Hadfield, *British Canals*, 243; Hadfield, *The Canal Age*, 163.

118. Peter Way, *Common Labour* (Cambridge: Cambridge University Press, 1997), 92–3.

119. John Belchem, *Merseypride: Essays in Liverpool Exceptionalism* (Liverpool: Liverpool University Press, 2000), 73.

120. M. R. Beames, "The Ribbon Societies: Lower Class Nationalism in Pre-Famine Ireland," in *Nationalism and Popular Protest in Ireland*, ed. C. H. E. Philpin (Cambridge: Cambridge University Press, 2002), 247.

121. Beames, "The Ribbon Societies," 253.

122. John Belchem, "Freedom and Friendship to Ireland: Ribbonism in Early 19th Century Liverpool," *International Review of Social History* 39 (1994): 33–56, 38.

123. Tom Garvin, "Defenders, Ribbonmen and Others: Underground Political Networks in Pre-Famine Ireland," in *Nationalism and Popular Protest in Ireland*, ed. C. H. E. Philpin (Cambridge: Cambridge University Press, 2002), 241.

124. Beames, "The Ribbon Societies," 253.

125. Belchem, "Freedom and Friendship to Ireland," 72.

126. Karl Marx, "Letter from Marx to Engels" (1859), in Karl Marx and Friedrich Engels, *Ireland and the Irish Question* (New York: Progress Publishers, 1971), 273.

127. Beames, "The Ribbon Societies," 263.

128. Eric Hobsbawm, *Primitive Rebels* (Manchester: Manchester University Press, 1971 [1959]), 171.

129. Eric Hobsbawm, *Bandits* (New York: Delacorte Press, 1969).

130. Belchem, "Freedom and Friendship to Ireland," 38.

131. Beames, "The Ribbon Societies," 248–249.

132. Ibid., 246.

133. Andrew Bourne, *Report of the Trial of Richard Jones* (Dublin: Hodges and Smith, 1840), 81.

134. Ibid., 83.

135. Ibid., 81.

136. Ibid., 84–85.

137. Ibid., 108.

138. Beames, "The Ribbon Societies," 246.

139. Parliament of Great Britain, *Hansard's Parliamentary Debates,* vol. 206 (London: Cornelius Buck, 1871), 7.

140. Ibid., 7.

141. Ibid., 1066.

142. Ibid., 7.

143. Douglas Francis, *Origins: Canadian History to Confederation* (Toronto: Nelson Education, 1992).

144. Tim Pat Coogan, *Wherever Green Is Worn: The Story of the Irish Diaspora* (Basingstoke, UK: Palgrave Macmillan, 2002).

145. Tom Ogorzalek, "Filibuster Vigilantly: Private Actors, the American State, and Territorial Expansion," presented at the Columbia Mini-APSA Conference, April 2008.

146. Way, *Common Labour*, 196.

147. Kevin Kenny, *Making Sense of the Molly Maguires* (New York: Oxford University Press, 1998), 10–11; see also Joseph Lee and Marion Casey, eds., *Making the Irish American: History and Heritage of the Irish in the United States* (New York: New York University Press, 2007), 374.

148. Friedrich Engels, "Engels to Eduard Bernstein, June 26, 1882" (1882), in Karl Marx and Friedrich Engels, *Ireland and the Irish Question* (New York: Progress Publishers, 1971), 334–5.

149. John Macdonald, *Troublous Times in Canada: A History of the Fenian Raids of 1866 and 1870* (Toronto: W. S. Johnston, 1910), 11.

150. Hereward Senior, *The Last Invasion of Canada* (Toronto: Dundurn, 1991), 64–65.

151. Karl Marx, *Capital: Volume I* (New York: International Publishers, 1967 [1867]), 666.

152. Commander Sweeny, quoted in Macdonald, *Troublous Times in Canada*, 14.

153. Macdonald, *Troublous Times in Canada*, 8.

154. See caption to plate in the April 7, 1866, issue of *Frank Leslie's Illustrated Newspaper*, 41.

155. *New York Times*, May 6, 1867, 4, and May 4, 1867, 5.

156. William J. Switala, *Underground Railroad in New Jersey and New York* (Mechanicsburg, PA: Stackpole Books, 2006), 84.

157. Macdonald, *Troublous Times in Canada*, 25.

158. Coogan, *Wherever Green Is Worn*, 392.

159. George Denison, *The Fenian Raid on Fort Erie* (Toronto: Rollo and Adam, 1866), 18.

160. Macdonald, *Troublous Times in Canada*, 28.

161. Ibid., 90.

162. Ibid., 89.

163. Alexander Somerville, *Narrative of the Fenian Invasion of Canada* (Hamilton, ON: Joseph Lyght, 1866), 126.

164. Macdonald, *Troublous Times in Canada*, 116.

165. Joshua Smith, *Borderland Smuggling: Patriots, Loyalists, and Illicit Trade in the Northeast, 1783–1820* (Gainesville: University of Florida Press, 2006).

166. Commander Sweeny, quoted in Macdonald, *Troublous Times in Canada*, 14.

167. Senior, *The Last Invasion of Canada*, 56.

168. Archives of the Irish Canadian Cultural Association of New Brunswick.

169. Smith, *Borderland Smuggling*, 37.

170. "Filibuster, n.," *OED Online* (Oxford University Press) (last accessed December 8, 2014).

171. Senior, *The Last Invasion of Canada*, 177–183.

172. J. E. Parsons, *West on the 49th Parallel* (New York: William Morrow, 1963), 17–18.

173. Wolfe, "Transportation and Politics," 186.

174. Archives Canada, Foreign Office Series 5, vol. 1506; and see Forrester, "Markers on the Forty-Ninth."

175. Archives Canada, Foreign Office Series 5, vol. 1506, 185–186.

176. Elliott Flower, "Smuggled Opium," *Pearson's Magazine* 21, no. 3 (1909): 323–329, 327.

177. Hippocrene Dictionaries, *Dutch-English, English-Dutch: With a Brief Introduction to Dutch Grammar* (New York: Hippocrene Books, 1990), 248.

178. See, for instance, the list of *dempingen*, or canal and waterfront infills, in Jeanine van Rooijen, *De drooglegging van Amsterdam: Een onderzoek naar gedempt stadswater* (Amsterdam: Bureau Monumentenzorg Amsterdam, 1996), 45–48.

179. Hippocrene Dictionaries, *Dutch-English, English-Dutch*, 248.

180. Karel Burggencate, *Engelsch Woordenboek: Tweede Deel* (Groningen: J. B. Wolters, 1901), 117.

181. *Sempervirens*, vol. 7 (Amsterdam: D. B. Centen, 1878), 371.

182. Antony Winkler Prins, *Winkler Prins' Geïllustreerde encyclopaedie I* (Amsterdam: Uitgevers-Maatshcappy, 1905), 490.

183. Diederik van Amstel, "Het Palingoproer," *De Nieuwe Gids* 2 (October 1886): 6. "Fiery sea" is *vuurzee* in the original Dutch.

184. Ronald van de Wal, *Of geweld zal worden gebruikt! Militaire bijstand bij de handhaving en het herstel van de openbare orde: 1840–1920* (Hilversum, The Netherlands: Verloren, 2003), 134–140.

Chapter 4: Chenangoes

1. Decaying car float stations are still (as of the early 2010s) visible at numerous locations along Manhattan's West Side waterfront, for instance at 60th Street and at 28th Street. Photo-documentation of New York Harbor's floating barge "graveyards" can be found in Thomas R. Flagg, *New York Harbor Railroads in Color*, vol. 1 (Scotch Plains, NJ: Morning Sun Books, 2000), 9.

2. Joshua Freeman, *Working-Class New York: Life and Labor since World War II* (New York: New Press, 2000), 7–8.

3. See discussion in Jacob Shell, "Innovation, Labor and Gridlock: The Unbuilt Freight Plan for Manhattan's Geography of Production," *Journal of Planning History* 9, no. 1 (2010): 3–20, esp. 4–7.

4. Robert Fitch, *The Assassination of New York* (New York: Verso Books, 1993), 106.

5. See, for instance, Edward Glaeser, *Triumph of the City: How Our Greatest Invention Makes Us Richer, Smarter, Greener, Healthier, and Happier* (New York: Penguin, 2011), 4–5.

6. James Hund, "The Bits and the Pieces," in *Made in New York*, ed. Max Hall (Cambridge, MA: Harvard University Press, 1959), 278–292.

7. Raymond Vernon, "International Investment and International Trade in the Product Cycle," *Quarterly Journal of Economics* 80, no. 2 (1966): 190–207.

8. Roy Helfgott, "New York's Dominance," in *Made in New York*, ed. Max Hall (Cambridge, MA: Harvard University Press, 1959), 47–66.

9. Fitch, *The Assassination of New York*, x–xxi.

10. Ibid., 197.

11. Ibid., 198.

12. Ibid., 203.

13. Ibid., 83. Fitch's portrayal of Moses's power to determine the built environment of New York contrasts sharply with the portrayal in Robert Caro's classic 1974 biography of Moses, *The Power Broker: Robert Moses and the Fall of New York* (New York: Knopf). Fitch portrays Moses as a mere instrument of the real estate and financial communities' wishes, whereas Caro portrays the financial and real estate communities as marginal to the actual planning of the city. The question regarding the true extent of Moses's power has been the subject of numerous other studies, including Joel Schwartz's 1993 *The New York Approach: Robert Moses, Urban Liberals and the Redevelopment of the Inner City* (Columbus: Ohio State University Press) and Hilary Ballon and Kenneth T. Jackson's 2008 edited volume, *Robert Moses and the Modern City: The Transformation of New York* (New York: W. W. Norton). Nevertheless, whatever the nature of collusion between the government authorities and the real estate and financial communities in mid-century New York, the fact is that *neither* ruling bloc ever pushed for manufacturing-friendly infrastructure. This in turn strongly suggests that certain attitudes and biases were influencing the thinking *both* of Moses *and* of the financial and real estate leaders with whom he colluded and occasionally clashed.

14. John Ensor Harr and Peter J. Johnson, *The Rockefeller Conscience* (New York: Charles Scribner's Sons, 1991), 392.

15. Fitch, *The Assassination of New York*, 57.

16. William Wilgus, *Proposed New Railway System for the Transportation and Distribution of Freight by Improved Methods in the City and Port of New York* (New York: Submitted to the Public Service Commission of the First District by the Amsterdam Corporation, 1908).

17. Ibid., 27.

18. Regional Plan Association, *Graphic Regional Plan*, vol. 2, *Building of the City* (New York: Committee of the Regional Plan, 1931), 395.

19. New York, New Jersey Port and Harbor Development Commission (PHDC), *Joint Report with Comprehensive Plan and Recommendations* (Albany, NY: J. B. Lyon Co., 1920).

20. Ibid., 279–280.

21. Ibid., 280.

22. Merchants Association of New York, *Industrial Map of New York City* (New York: Merchants Association of New York, 1922).

23. PHDC, *Joint Report with Comprehensive Plan*, 1.

24. Letter, Port of New York Authority to Julius Miller, Feb. 6, 1925, New York Public Library pamphlet TLC p.v. 227 no. 1, 12.

25. Benedict Crowell and Robert Forrest Wilson, *How America Went to War: The Road to France Part I* (New Haven, CT: Yale University Press, 1921), 113. See also Josef Konvitz, "The Crisis of Atlantic Port Cities, 1880 to 1920," *Comparative Studies in Society and History* 36, no. 2 (1994): 293–318. And see PHDC, *Joint Report with Comprehensive Plan*, 41–61.

26. See PHDC, *Joint Report with Comprehensive Plan*, 50–57. See also Jameson Doig, *Empire on the Hudson: Entrepreneurial Vision and Political Power at the Port of New York Authority* (New York: Columbia University Press, 2001), 27–45.

27. Erwin Bard, *The Port of New York Authority* (New York: Columbia University Press, 1942), 6–10.

28. See *New York Times*, Aug. 30, 1924, 4, and July 29, 1921, 20.

29. Regional Plan Association, *Graphic Regional Plan,* vol. 2, 339.

30. Letter, Port of New York Authority to Miller (1925), 17.

31. See Frederick Bird, *A Study of the Port of New York Authority* (New York: Dun and Bradstreet, 1948), 101–102. See also Port of New York Authority, *Container Shipping: Full Ahead* (New York: Port of New York Authority, 1967).

32. Doig, *Empire on the Hudson*, 143–179.

33. Lusk Committee Papers, Box 2, Folder 27, New York State Archives.

34. See, for instance, the American Coalition of Patriotic, Civil and Fraternal Societies' *Map of the Borough of Manhattan Showing Relation between Distribution of Foreign Stock and Negro Population with the Dissemination of Communist and Socialist Propaganda during the Summer of 1932,* Tamiment Library and Robert F. Wagner Labor Archives, New York University.

35. Robert F. Zeidel, "Pursuit of 'Human Brotherhood': Philanthropy and American Immigration, 1900-1933," *New York History* 90, nos. 1–2 (2009), 85–106.

36. Beverly Gage, *The Day Wall Street Exploded: A Story of America in Its First Age of Terror* (New York: Oxford University Press, 2009), 112.

37. Ibid., 190–191.

38. Thomas J. Tunney, *Throttled! The Detection of the German and Anarchist Bomb Plotters,* as told to Paul Merrick Hollister (Boston: Small, Maynard and Co., 1919), 11.

39. Michael Hadley and Roger Sarty, *Tin-pots and Pirate Ships: Canadian Naval Forces and German Sea Raiders, 1880–1918* (Montreal: McGill University Press, 1991), 111.

40. Tunney, *Throttled!*, 128.

41. Carlo Tresca, *The Autobiography of Carlo Tresca* (New York: John D. Calandra Italian American Institute, Queens College, City University of New York,, 2003), 252–253; Tunney, *Throttled!*, 40, 59–60.

42. Maia Ramnath, *Haj to Utopia: How the Ghadar Movement Charted Global Radicalism and Attempted to Overthrow the British Empire* (Berkeley: University of California Press, 2011); Tunney, *Throttled!*, 103.

43. Tunney, *Throttled!*, 77.

44. Ibid., 253.

45. Ibid., 264.

46. Ibid., 249.

47. Ibid., 251.

48. RKO Radio Pictures, "Manhattan Waterfront," *World on Parade Series* (Van Beuren Corp., 1937). And see Madeline Dickens, "A Floating Village at Anchor," *Everywhere* 28 (1911): 274.

49. Tunney, *Throttled!*, 127.

50. William Shakespeare, *The Merchant of Venice* I.iii.15–23, in *The Riverside Shakespeare*, ed. Blakemore Evans (New York: Houghton Mifflin, 1974 [1598]), 258.

51. Tunney, *Throttled!*, 131–133.

52. Ibid., 132.

53. Ibid., 134.

54. Ibid., 139.

55. Ibid., 139–140.

56. Ibid., 172.

57. Ibid., 140–141.

58. Ibid., 148.

59. Ibid., 149.

60. Ibid., 137.

61. Ibid., 138.

62. Ibid., 139.

63. Ibid., 159.

64. Ibid., 176–177.

65. Charles Barnes, *The Longshoremen* (New York: Russell Sage Foundation, 1915), 284.

66. Robert Hendrickson, *Salty Words* (New York: Hearst Marine Books, 1984), 119, and Charles Cutler, *O Brave New Words! Native American Loanwords in Current English* (Norman: University of Oklahoma Press, 2000), 198.

67. Peter Way, *Common Labour: Workers and the Digging of North American Canals, 1780–1860* (Cambridge: Cambridge University Press, 1993); and see Kevin Kenny, *Making Sense of the Molly Maguires* (New York: Oxford University Press, 1998).

68. Joel T. Headley, *The Great Riots of New York 1712–1873* (New York: Cosimo, 2009 [1873]), 280–300.

69. Arthur Wyllie, *The Union Navy* (Raleigh, NC: Lulu Press, 2007).

70. Milton F. Perry, *Infernal Machines: The Story of Confederate Submarine and Mine Warfare* (Baton Rouge: Louisiana State University Press, 1965), 136.

71. Royal Bird Bradford, *History of Torpedo Warfare* (Newport, RI: U.S. Navy, 1882), 62.

72. Philip von Doren Stern, *Secret Missions of the Civil War: First-hand Accounts* (Chicago: Rand McNally, 1959), 261.

73. "E," letter, *London Times*, Dec. 28, 1875, and *New York Daily Herald*, Jan. 17, 1876.

74. Ann Larabee, *The Dynamite Fiend: The Chilling Story of Alexander Keith Jr., Nova Scotian Spy, Con Artist, and International Terrorist* (New York: Palgrave Macmillan, 2005).

75. "E," letter.

76. Ibid.

77. Ibid.

78. *National Quarterly Review* editors, "Fortified Cities," *National Quarterly Review* 24 (1871): 86–110, 99.

79. Ibid., 101.

80. Gage, *The Day Wall Street Exploded*, 46.

81. Ibid., 42.

82. Johann Most, *Revolutionäre Kriegswissenschaft: Ein Handbüchleinzur Anleitung betreffend Gebrauches und Herstellung von Nitro-Glycerin, Dynamit, Schiessbaumwolle, Knallquecksilber, Bomben, Brandsätzen, Giften, u.s.w., u.s.w.* (New York: Druck und Verlag des Internationalen Zeitungs-Vereins, 1885).

83. David Clarke, *Technology and Terrorism* (New Brunswick, NJ: Transaction Publishers, 2004), 39.

84. Samuel Gompers, *Seventy Years of Life and Labor* (New York: E. P. Dutton, 1925), 174, 215.

85. Gage, *The Day Wall Street Exploded*, 64; and see Sven Beckert, *The Monied Metropolis: New York City and the Consolidation of the American Bourgeoisie* (Cambridge: Cambridge University Press, 2003), 1–2.

86. Paul Avrich, *Sacco and Vanzetti: The Anarchist Background* (Princeton, NJ: Princeton University Press, 1996), 100.

87. "Intruder Has Dynamite," *New York Times,* July 4, 1915.

88. Tunney, *Throttled!,* 49.

89. "Bomb Rocks Police Headquarters," *New York Times,* July 6, 1916, 105.

90. Harry Landau, *The Enemy Within* (New York: G. P. Putnam's Sons, 1937), 140.

91. Jules Witcover, *Sabotage at Black Tom: Imperial Germany's Secret War in America, 1914–1917* (Chapel Hill, NC: Algonquin Books of Chapel Hill, 1989), 12.

92. Gage, *The Day Wall Street Exploded,* 130–135.

93. Charles Howard McCormick, *Hopeless Cases: The Hunt for the Red Scare Terrorist Bombers* (Lanham, MD: University Press of America, 2005), 86.

94. Luigi Galleani, *La salute e in voi!,* dated 1905, held in the collection of the International Institute of Social History, Amsterdam, The Netherlands.

95. Gage, *The Day Wall Street Exploded,* 215.

96. Robert Fishman, "The Regional Plan and the Transformation of the Industrial Metropolis," in *The Landscape of Modernity: New York 1900–1940,* ed. David Ward and Olivier Zunz (Baltimore, MD: Johns Hopkins University Press, 1997).

97. Fitch, *The Assassination of New York,* 56–60.

98. Barnes, *The Longshoremen,* 7.

99. Ibid., 14.

100. Ibid., 14–15.

101. Ibid., 93.

102. Ibid., 18.

103. Ibid., 15.

104. Ibid., 16.

105. Ibid., 41.

106. Waldo Browne, *What's What in the Labor Movement: A Dictionary of Labor Affairs and Labor Terminology* (New York: W. B. Heubsch, 1921), 59.

107. See, for example, "Shenangoes Sue Waterfront Unit," *New York Times,* Sept. 1, 1955, 46.

108. See, for example, Vernon Jensen, "Decasualization of Employment on the New York Waterfront," *Industrial and Labor Relations Review* 11, no. 4 (1958): 534–550, 540.

109. Regional Plan Association, *Graphic Regional Plan,* vol. 2, 364.

110. Robert Murray Haig, *Regional Survey of New York and Its Environs,* vol. 1, *Major Economic Factors in Metropolitan Growth and Arrangement* (New York: Russell Sage Foundation, 1927), 32–33.

111. Ibid., 9.

112. Ibid., 23.

113. Ibid., 32.

114. Ibid., 33.

115. Ibid., 44.

116. Ibid., 41.

117. Robert Murray Haig, *Regional Survey of New York and Its Environs,* vol. 1A, *Chemical, Metal, Wood, Tobacco and Printing Industries* (New York: Russell Sage Foundation, 1928), 8.

118. Haig buries the fact of this absolute growth elsewhere in the survey and downplays the trend's importance, on the dubious grounds that absolute rates of growth among chemical factories in the outskirts of the region have been higher: "While it is true that in the very heart of the city the number of plants has increased about ten per cent and the number of employees about forty per cent during the 22 year period, this growth is insignificant as compared with the enormous gains recorded in all of the outlying sections" (ibid., vol. 1A, 18–19).

119. Haig, *Regional Survey of New York and Its Environs,* vol. 1, 100.

120. Haig, *Regional Survey of New York and Its Environs,* vol. 1A, 13–16, 29.

121. Hund, "The Bits and the Pieces," 278–292.

122. Robert Bothwell, *Eldorado: Canada's National Uranium Company* (Toronto: University of Toronto Press, 1984), 79–116.

123. See chapter 3, this work.

124. John Donaldson, *A Canoe Quest in the Wake of Canada's "Prince of Explorers"* (Kingston, ON: Artful Codger, 2006), 262. See also Peter van Wyck, "The Highway of the Atom: Recollections along a Route," *Topia* 7 (2006): 99–115.

125. Richard Hewlett and Oscar E. Anderson, *The New World: A History of the United States Atomic Energy Commission, 1939–1946* (University Park: Pennsylvania State University Press, 1962), 86. See also Van Wyck, "The Highway of the Atom," 109. And see Cynthia Kelly and Robert Norris, *A Guide to Manhattan Project Sites in Manhattan* (Washington, DC: Atomic Heritage Foundation, 2008), 4–8.

126. U.S. Government Department of Energy, *Memorandum: Authorization for Remedial Action at the Former Baker-Williams Warehouses on West 20th Street in New York, New York*, DOE F1325–8 (8–89) (Washington, DC: Government Printing Office, 1990). See also William Broad, "Why They Called It the Manhattan Project," *New York Times,* Oct. 30, 2007.

127. U.S. Government Department of Energy, *Memorandum*, 6, 28, 39.

128. Hewlett and Anderson, *The New World*, 86.

129. Van Wyck, "The Highway of the Atom."

130. "State Department Involved in Spy Case and Interference with FBI," *National Republic* 33, no. 11 (1946): 9. The *National Republic* article refers to an investigative report in the *Montreal Gazette*, which can be found on p. 1 of the February 19, 1946, issue of that paper. The House Un-American Activities Commission's interest in the Canadian black market uranium case is discussed in George Racey Jordan, *From Major Jordan's Diaries* (New York: Harcourt, Brace and Co., 1952), 61.

131. Joseph Slater, "Labor and the Boston Police Strike of 1919," in *The Encyclopedia of Strikes in American History*. ed. Aaron Brenner, Benjamin Day, and Immanuel Ness (Armonk, NY: M. E. Sharpe, 2009), 239–251.

132. See the section on Meyer London in Cohen's *They Built Better Than They Knew* (New York: Ayer Publishing, 1946), 215–222.

133. See Cohen's 1916 *Law and Order in Industry* (New York: Macmillan Press), his 1918 *Commercial Arbitration and the Law* (New York: D. Appleton and Co.), his 1919 *An American Labor Policy* (New York: Macmillan Press), and his 1920 article, "Collective Bargaining and the Law as a Basis for Industrial Reorganization," *Annals of the Academy of Political and Social Science* 90:47–49.

Conclusion

1. James C. Scott, *Seeing Like a State: How Certain Schemes to Improve the Human Condition Have Failed* (New Haven, CT: Yale University Press, 1998), 2.

2. James Vance, *Capturing the Horizon: The Historical Geography of Transportation since the Transportation Revolution of the Sixteenth Century* (New York: Harper and Row, 1986), 2.

Index